VOLUME ONE HUNDRED AND TWENTY FOUR

# PROGRESS IN MOLECULAR BIOLOGY AND TRANSLATIONAL SCIENCE

## Genetics of Cardiovascular Disease

VOLUME ONE HUNDRED AND TWENTY FOUR

# PROGRESS IN MOLECULAR BIOLOGY AND TRANSLATIONAL SCIENCE

## Genetics of Cardiovascular Disease


Edited by

**TIMOTHY J.A. CHICO**

*MRC Centre for Developmental and Biomedical Genetics, and Department of Cardiovascular Science, University of Sheffield, Sheffield, United Kingdom*



ELSEVIER

AMSTERDAM • BOSTON • HEIDELBERG • LONDON
NEW YORK • OXFORD • PARIS • SAN DIEGO
SAN FRANCISCO • SINGAPORE • SYDNEY • TOKYO

Academic Press is an imprint of Elsevier

Academic Press is an imprint of Elsevier
225 Wyman Street, Waltham, MA 02451, USA
525 B Street, Suite 1800, San Diego, CA 92101-4495, USA
Radarweg 29, PO Box 211, 1000 AE Amsterdam, The Netherlands
The Boulevard, Langford Lane, Kidlington, Oxford, OX5 1GB, UK
32 Jamestown Road, London NW1 7BY, UK

First edition 2014



**Library of Congress Cataloging-in-Publication Data**
A catalog record for this book is available from the Library of Congress

**British Library Cataloguing in Publication Data**
A catalogue record for this book is available from the British Library

ISBN: 978-0-12-386930-2
ISSN: 1877-1173

For information on all Academic Press publications
visit our website at store.elsevier.com

# CONTENTS

# CONTRIBUTORS

**Bill Chaudhry**
Institute of Genetic Medicine, Newcastle University, Newcastle, United Kingdom

**Timothy J.A. Chico**
MRC Centre for Developmental and Biomedical Genetics, and Department of Cardiovascular Science, University of Sheffield, Sheffield, United Kingdom

**Sheila E. Francis**
Department of Cardiovascular Science, Medical School, University of Sheffield, Sheffield, United Kingdom

**Deborah J. Henderson**
Institute of Genetic Medicine, Newcastle University, Newcastle, United Kingdom

**Chris Jopling**
CNRS, UMR-5203, Institut de Génomique Fonctionnelle, Département de Physiologie, Labex Ion Channel Science and Therapeutics; INSERM, U661, and Universités de Montpellier 1&2, UMR-5203, Montpellier, France

**Allan Lawrie**
Department of Cardiovascular Science, University of Sheffield, Sheffield, United Kingdom

**Peter Novodvorsky**
MRC Centre for Developmental and Biomedical Genetics, University of Sheffield, Firth Court, and Department of Cardiovascular Sciences, University of Sheffield, Sheffield, United Kingdom

**Simon Ramsbottom**
Institute of Genetic Medicine, Newcastle University, Newcastle, United Kingdom

**Alex M.K. Rothman**
Department of Cardiovascular Science, University of Sheffield, Sheffield, United Kingdom

**Robert F. Storey**
Department of Cardiovascular Science, University of Sheffield, Sheffield, United Kingdom

**Mark R. Thomas**
Department of Cardiovascular Science, University of Sheffield, Sheffield, United Kingdom

**Fredericus J.M. van Eeden**
MRC Centre for Biomedical Genetics, Department of Biomedical Science, University of Sheffield, Sheffield, United Kingdom

**Robert N. Wilkinson**
Department of Cardiovascular Science, Medical School, University of Sheffield, Sheffield, United Kingdom

# PREFACE

It is customary, though unnecessary, to preface a volume such as this by highlighting the burden of cardiovascular disease. Suffice it to say that the majority of this volume's readers (and contributors) will develop some form of cardiovascular disease during their lifetime. I look forward to a time when publications such as this will be of historic interest only and intend to live to see it, if possible.

The ongoing battle against cardiovascular disease should not detract from the huge successes made in the last century. We now understand more about the causes of cardiovascular disease than could have been imagined a few decades ago. With this has come a range of diagnostics and treatments that have radically improved our ability to identify, treat, or prevent such diseases. However, there is no room for complacency; every hospital bed contains an example of our failure to fully understand and address why people get sick.

The influence of the genome and its regulation on cardiovascular disease has long been suspected. On occasion, debate has ensued regarding the relative contributions of heritability versus environmental and societal factors to an individual's risk of developing cardiovascular disease. My personal view is that no real distinction between such pressures exists; at some level every element of human life is determined by the sum total of the genomes of many millions of individuals, including those of previous generations. What I'm trying to say is that, basically, it's all about the genes.

The futility of attempting to comprehensively cover the genetics of cardiovascular disease is obvious, and I have made no such attempt. However, this volume assembles a range of reviews that present accessible yet comprehensive descriptions of areas of current interest in cardiovascular science. Each would educate and inform all but experts in the respective areas; I have certainly learnt a great deal editing them.

Chapter 1 by Sheila Francis is extremely timely, coming at a time when some angst is being voiced about whether the ubiquitous mouse models of atherosclerosis are sufficiently informative about human disease. It remains to be seen which way the field will go, and it is to be hoped that the difficult decisions faced by research funders will be the right ones.

Chapter 2 by Bill Chaudhry and Deborah Henderson presents a summary of the genetics of mammalian cardiac development. The field of developmental biology (formerly embryology) has been rejuvenated by the prospect that developmental processes underpin many, if not most, diseases; and that recapitulation of such phenomena by "regenerative medicine" holds the prospect of a truly novel approach to treating disease. We will know in the next two decades whether this prospect will flourish or founder.

Chapter 3 by Alex Rothman, myself, and Allan Lawrie presents a study of the contribution of microRNA to pulmonary vascular disease. The focus on a rare though devastating disease should not prevent this chapter being of interest to all cardiovascular researchers, since the importance of microRNAs and the promise of therapeutic manipulation extend to all areas of human disease. The paradigm shift provoked by the discovery of microRNAs only a few years ago has already led to plausible strategies for treatment of cardiovascular disease, and it is my prediction that patients with cardiovascular diseases will be treated by these approaches before this volume is entirely outdated.

Chapters 4 and 5 highlight a new player in the field of cardiovascular genetics and a pet shop favorite, the zebrafish. Robert Wilkinson, Chris Jopling, and Fredericus van Eeden present in Chapter 4 a summary of the use of zebrafish to model cardiac disease, while in Chapter 5 Wilkinson and van Eeden outline how the same humble organism can be used to understand vascular development and disease. The advantages of this model have led to fundamental breakthroughs in our understanding of cardiovascular development. Scientists less familiar with the system are encouraged to consider how it could be applied to their own areas of interest.

Chapter 6 by Mark Thomas and Robert Storey describes the cutting-edge interface between genetics and the response to antiplatelet drugs, one of the hugely successful therapies alluded to above. I will allow myself another prediction that the next decade will see a move toward more personalized drug prescription and that antiplatelet therapy, with both clear benefits and significant risks, will be in the vanguard of this approach.

Chapter 7 by Peter Novodvorsky and myself describes the transcription factor KLF2 and its contribution to vascular development and disease. Another major success story, the HMG-CoA reductase inhibitors (statins) may owe some of their success to this molecule, and approaches to manipulate KLF2 might lead to significant benefits for a range of cardiovascular diseases.

Until my involvement in this volume I had not fully understood why prefaces contained expressions of gratitude; I now realize these are veiled apologies. With this in mind, I would like to thank Mary-Ann Zimmerman, Helene Kabes, and the production team at Elsevier for their patience and hard work. I would also like to thank the contributors; where this volume succeeds, it is by their efforts. The responsibility for any deficiencies is mine alone.

TIMOTHY J.A. CHICO
Sheffield, UK

CHAPTER ONE

# Genetic Experimental Preparations for Studying Atherosclerosis

**Sheila E. Francis**
Department of Cardiovascular Science, Medical School, University of Sheffield, Sheffield, United Kingdom

## Contents



## Abstract

Atherosclerosis is a pathological process with several inputs (biological, chemical, physiological, and others) interacting slowly over a lifetime leading to coronary artery disease, significant morbidity, and a limited lifespan. Over the past two decades, biologists have used experimental preparations from cells, animals, and man to understand the biology of atherosclerosis. Much has been discovered but our use of the standard gene-targeted experimental preparations is now nearing its limit. Better preparations to answer the remaining questions in the field of atherosclerosis biology are needed.

*Progress in Molecular Biology and Translational Science*, Volume 124
ISSN 1877-1173
http://dx.doi.org/10.1016/B978-0-12-386930-2.00001-X

## 1. INTRODUCTION

Much is already known about the complex and multifaceted process of atherosclerosis from research using genetic experimental systems in whole animals. Progress has been significant on the biology of atherosclerotic lesions, and some studies have led directly to treatments in man. Despite this, there remains an excess of treatable coronary artery disease (15%, some studies suggest) despite the best modern therapies.[1] There is a need therefore to discover new molecular and cellular mechanisms, and this is being addressed using approaches such as genome-wide association studies (GWAS). Currently over 30 of these studies have been published, the first being in 2007[2] with CARDIoGRAM[3] being the most recent. These gene discoveries do need to be evaluated using reliable experimental preparations, and increasingly, these experimental systems are expected to test therapeutic modalities/devices as well. With increased focus upon translational science, it is important that genetic preparations of atherosclerosis are as close as possible to the situation in humans that leads to presentation of the symptoms of advanced atherosclerosis, for example, angina or myocardial infarction. This chapter reviews the currently used preparations in animals with a specific focus on new developments (gene-targeted rats and pigs) and experimental approaches (bone marrow transplantation studies) that could lead up to new treatments in man.

## 2. WHY WE NEED MORE AND BETTER WAYS TO STUDY ATHEROSCLEROSIS

Although the seminal studies by Michael Davies,[4] Renu Virmani,[5] and others have provided valuable insights into the pathogenesis of human atherosclerosis, in order to evaluate the new targets produced by unbiased gene discovery approaches, it remains necessary to study *in vivo* models where individual genes are deleted or misregulated to examine their biological consequences. The most common approach has been to use embryo-directed targeting such that animals (e.g., mice) lack a particular gene from embryonic development throughout their life. This method has been dictated by the progress in this area of technology and while it produces a "clean" system/preparation for study, it does not take into account latent effects caused by deletion in early life nor compensations that occur as a result of the deletion. Research so far has produced many fascinating insights into the biology of atherosclerosis but still the molecules that, for example, lead to plaque rupture remain elusive.

Newer systems of gene deletion using cre–lox technologies[6] allow deletion of the function of a gene at specific times and in selected cell types, for example, when the mice are adults and in smooth muscle cells, for example. Increasingly, more is being expected of the genetic preparations of atherosclerosis in terms of their ability to mimic all facets of human disease, for example, plaque rupture and plaque repair. As the field progresses further and the clinical translational challenges become even greater, *in vivo* preparations need to be versatile enough to allow for exploration of these newer challenges.

Indeed, the field should try to move away from using the loose term "model" in descriptions of whole-animal studies of atherosclerosis as it is clear that none of the current so-called models, perhaps with the exception of the proprotein convertase subtilisin/kexin type 9 (PCSK9) transgenic pig, really mimic the clinical presentation of coronary artery disease as in humans. I prefer the term experimental preparations since this implies mimicry of the biological processes and not the entire disease phenotype. The field needs to consider developing new or more sophisticated research tools such as computer simulations of atherosclerosis, for example, digital atherosclerotic mice and new large animal preparations.

## 2.1. Gene-targeted mouse models of atherosclerosis

Although there is variation among the strains, generally, wild-type mice are not susceptible to atherosclerosis. Therefore, genetic approaches have been developed to engineer mice that develop fatty plaques in their aorta. The approaches used so far have targeted genes responsible for cholesterol uptake and these lead to raised plasma cholesterol, in some cases even when fed a regular rodent chow diet. The most widely used *in vivo* mouse models are detailed in Table 1.1.

Feeding of high-fat diets to gene-targeted mice (Table 1.1) allows researchers to study the biology of atherosclerosis in a time frame of 8–26 weeks. The main diets used have been reviewed elsewhere[12] and these contain 20% (w/v) fat (derived from lard or cocoa butter), called Western-type diets, and these can be supplemented by other constituents, for example, emulsifiers like cholic acid. Cholic acid promotes cholesterol absorption, inhibits conversion to bile acids, reduces clearance, and results in abnormally high concentrations of cholesterol in the plasma. It has also been reported that cholate modifies inflammatory gene expression.[13]

**Table 1.1** Genetic experimental preparations of atherosclerosis in mice

| | Strain background | Diet required | Reference |
|---|---|---|---|
| $ApoE^{-/-}$ | C57BL/6 | Spontaneous development of lesions, accelerated by a high-cholesterol diet Hyperlipidemia > 40 mM cholesterol, main lipid particles are VLDL | 7,8 |
| $LDLR^{-/-}$ | C57BL/6 | Requires a fatty diet to develop lesions. A milder model than $ApoE^{-/-}$ | 9 |
| $LDLR^{-/-}bec1^{-/-}$ | C57BL/6 | Spontaneous lesion development accelerated by diet. Males > females (lesion area of aorta) Hyperlipidemia > 40 mM cholesterol on a HF diet | 10 |
| $LDLR^{-/-}ApoE^{-/-}$ | C57BL/6J, 129Sv | Spontaneous elevated cholesterol, not higher in DKO than $ApoE^{-/-}$ alone, approximately 15 mM | 11 |

### 2.2. Gender effects on experimental preparations of atherosclerosis

Most but not all studies on the biology of atherosclerosis in whole animals use males. This is most often to avoid estrogen effects on biological processes as some genes are transcriptionally modulated by estrogen as a result of an estrogen response element in their promoter.[14] Atherosclerosis in mice is affected by estrogen.[15] Researchers also note that effects of interventions upon the development of atherosclerosis in male cohorts are more consistent leading to manageable group sizes in experimental studies although in practice atherosclerotic lesions are largest in females.[16,17]

## 3. APOLIPOPROTEIN E KNOCKOUT MICE FOR STUDY OF ATHEROSCLEROSIS BIOLOGY

This strain is one the most widely used in the laboratories worldwide for the study of processes involved in the development of atherosclerosis. Homozygous deficiency of apolipoprotein E in mice from birth[7,8] leads to hyperlipidemia with VLDL being the most common particle to be elevated in the serum of these animals. Atherosclerosis ensues[18] and is even present in ApoE$^{/-}$ heterozygous mice.[19] ApoE is an important modulator of lipoprotein interactions with several receptors responsible for lipid clearance.[20] Although hyperlipidemia occurs spontaneously, this can be accelerated by feeding a high-fat diet and cholesterol concentrations in the blood can rise to 40 mM or higher.[7] The progress of atherosclerosis in ApoE$^{-/-}$ mice is well documented. Lesions arise in the aorta and in the great vessels and become complex in nature (foam cells, increased collagen, and plaques) as a result of fat feeding.[21,22] The location of plaques in mice contrasts with that in humans where lesions are found in the coronaries, the carotids, and the cerebral vessels. There is a high prevalence of lesions at the ostium and the region immediately surrounding which does not replicate the human pattern of disease.[23] Intraplaque hemorrhage has been detected in lesions and is thought by some to be a sign of previous plaque rupture and healing.[22]

## 4. LOW-DENSITY LIPOPROTEIN KNOCKOUT (LDLR$^{-/-}$) MICE FOR STUDY OF THE BIOLOGY OF ATHEROSCLEROSIS

LDLR$^{-/-}$ deficiency in humans leads to hyperlipoproteinemia and without treatment individuals suffer a premature myocardial infarction.[24] Mice lacking LDLR do not develop atherosclerosis spontaneously but need to be fed a high-fat diet in order to develop substantial plaques in their aorta.[9] The feeding period is usually slightly longer than that used to generate experimental lesions in ApoE$^{-/-}$ mice. Once the feeding period is complete, lesions generally achieve the same degree of complexity as those in ApoE$^{-/-}$ mice. It is noteworthy that the plasma concentrations of cholesterol are more akin to the human condition in terms of degree and the distribution of lipid species produced. This is due to the presence of a truncated form of ApoB (called B48) on LDL particles synthesized by the liver in these mice which permits enhanced binding of ApoE and promotes other non-LDLR-related clearance mechanisms.[25]

## 5. MIXED GENE-TARGETED MANIPULATIONS FOR THE STUDY OF ATHEROSCLEROSIS

The most well known of these is apobEC-1$^{-/-}$ LDLR$^{-/-}$, and these mice are used when investigators wish to specifically increase LDL cholesterol in their experimental system. ApobEC-1 is an RNA-editing enzyme[26] which gives rise to a truncated version of apoB100. If this is deleted in mice, the full-length form of apoB100 persists (rather than the truncated apoB48 form) and this is more akin to humans, where apoB100 predominates. For the study of atherosclerosis, apobEC-1$^{-/-}$ mice need to be combined with LDLR$^{-/-}$ to generate the correct atherosclerosis preparation for study.[10]

### 5.1. Bone marrow transplantation as a tool for the study of atherosclerosis

This technique is now used quite widely in mouse and rat atherosclerosis preparations and is useful for providing insight into any difference between the effects of a protein in the vessel wall versus that in haematopoietic cells.[27] A distinct advantage of the technique is that it allows for manipulations in adult mice rather than including time elapsed from embryogenesis onward. It has gained popularity alongside the cre–lox system as an additional tool in the researchers' armamentarium for studies of experimental atherosclerosis. Data gained using this technique can reveal opportunities for intervention in hematopoietic cells. One of the first studies in atherosclerosis using this technique revealed a role for platelet P-selectin in addition to that already known for endothelium.[28] There are some methodological issues that need to be taken into account in atherosclerosis studies that incorporate bone marrow depletion and reconstitution, one being that irradiation itself has a negative effect upon development of atherosclerotic lesions. For this reason, the correct transplantation controls should always be performed in these types of experiments.

## 6. METHODS USED TO STUDY ATHEROSCLEROTIC PREPARATIONS

Numerous methods are used to study atherosclerosis and the range of approaches is ever increasing. In the main, however, researchers are interested in the size of the lesions and in their constituents at cellular and molecular levels. All researchers usually examine the surface of the aorta, called the *en face* approach. After perfusion fixation, entire lengths of aorta are carefully dissected and opened out longitudinally. Aortas are cleaned, pinned onto a surface (usually wax), and stained with Oil red O or Sudan IV to identify areas of the intima containing lipid lesions (Fig. 1.1A and Ref. 29).

In mice, two other locations are commonly studied, the aortic root and the innominate (brachiocephalic) artery. The aortic root, a specific region of

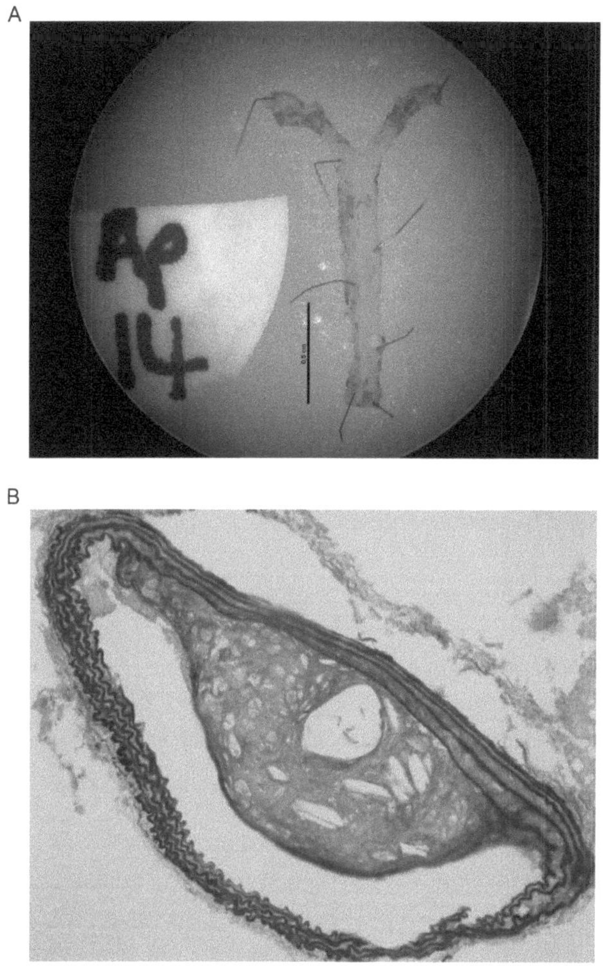


**Figure 1.1** Histological outputs from experimental atherosclerosis (ApoE$^{-/-}$ fed a Western diet) in mice. (A) An *en face* preparation of aorta stained with Oil red O to show lipid deposits, scale bar 0.5 cm. (B) A cross section of the brachiocephalic artery with Elastic van Gieson staining to show an eccentric plaque, artery width at its widest part is 0.5 mm. (See the color plate.)

the ascending aorta 400 μm above the coronary ostia, remains the most common area for analysis. There is increasing emphasis on lesions in the innominate artery, since this is described as being a site where nonfatal plaque rupture leads to buried fibrous caps and intraplaque hemorrhages (Fig. 1.1B and Refs. 22,30).

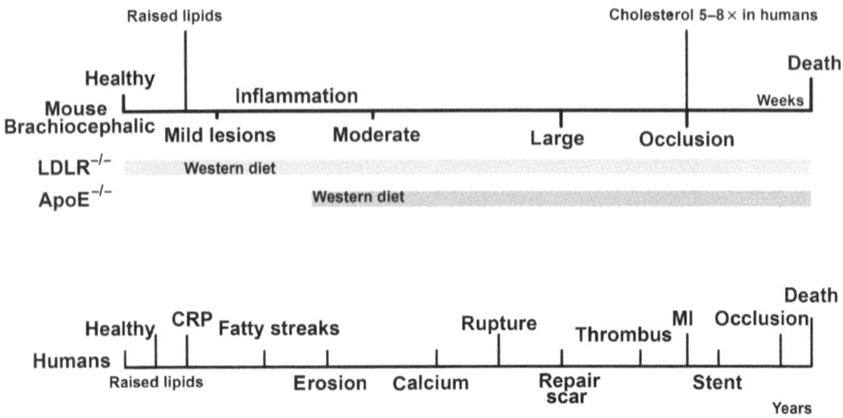


**Figure 1.2** A schematic to show the main differences in pathology between the experimental mouse preparations and the process of human atherosclerosis. Top: relatively few phases of atherosclerosis in the experimental preparations. Bottom: the complexity of the human pathology of atherosclerosis. (See the color plate.)

After perfusion fixation, tissue is usually embedded in paraffin wax and thin cross-sectional slices cut on a microtome and placed onto coated glass slides before histological staining.

There are numerous other methods in *ad hoc* use for the longitudinal study of atherosclerosis. There are various imaging modalities used in the clinical arena and adapted for small animal preparations, for example, echocardiography, MR (magnetic resonance),[31,32] positron emission tomography,[33] etc. In MR imaging of mice with atherosclerosis, plaque components are distinguished by their different inherent signal intensities compared with the main components of the vessel wall. This research area is challenging due to limits on spatial resolution of the various techniques being tried (not just MR but others also) but this is an area of active research. Figure 1.2 is a schematic contrasting the animal and human preparations of atherosclerosis in terms of their disease pathology.

## 7. ZEBRAFISH PREPARATIONS FOR THE STUDY OF ATHEROSCLEROSIS

Zebrafish have almost unlimited potential for genetic modifications, and a handful of studies have now shown that zebrafish fed a high-cholesterol diet exhibit lesions resembling early atherosclerosis in humans.[34] In this preparation, accumulation of LDL can be visualized in vessels and this is decreased when the fish are switched to a regular diet or exposed to antioxidant agents. This system may be useful when testing therapies for atherosclerosis especially where lipid accumulation is the primary pathological process under investigation.

## 8. GENETICALLY TARGETED RAT PREPARATIONS FOR THE STUDY OF ATHEROSCLEROSIS

There are a number of rat preparations developed via commercial organizations and the KO Rat Consortium including APOE, LDLR, SOD3, and P53. The APOE KO rat is on a Sprague–Dawley background and arises as a result of a 16 bp deletion within exon 3 of the *APOE* gene on chromosome 1. Homozygous KO rats produce no APOE protein detectable by Western blot. Total cholesterol concentrations are elevated fivefold to approximately 5–600 mg/dl in males and females. The administration of high-fat diets to APOE KO rats has led to a significantly reduced lifespan, and there are no published papers that use the APOE KO rat for the study of atherosclerosis. Similarly, the *LDLR* KO rat (Sprague–Dawley background) has a 337 bp deletion and a 4 bp insertion within exon 4 of *Ldlr* on chromosome 8. On regular chow, knockouts have a greater body weight and a four- and fivefold elevation of serum cholesterol compared to wild types or heterozygotes. These rats can tolerate a high-fat diet which further raises cholesterol by 50%. There appear to be no published studies with the proatherosclerotic strains to date but data are arising in other areas, for example, hypertension.[35]

## 9. GENETICALLY TARGETED PORCINE PREPARATIONS FOR THE STUDY OF ATHEROSCLEROSIS

Although the mouse has been useful in the study of the biology of atherosclerosis, several challenges still exist. For example, reliable imaging biomarkers to identify individuals at risk or their use where treatments are to be evaluated are not easy to validate in rodents. Near to human-sized vessels, preferably with atherosclerosis are needed to facilitate further technical development that will lead to new advances. There is also a need to study late-stage atherosclerosis including atherosclerotic plaque rupture, since this has not been satisfactorily tackled in rodent models, although has been tried.[30]

In addition, in rodents, due to the small size of the vessels studied, it has not really been possible to research new treatment approaches such as intravascular devices, for example, bioresorbable stents and newer imaging modalities such as optical coherence tomography. In addition to this, many important features of human plaques are absent in mice but present in porcine preparations, for example, preexisting intima (adaptive intimal thickening), thrombosis, and angiogenesis. The near-human size, genomic advances, and physiological and anatomical features make the porcine preparation the one of choice for just prior to man studies.[36,37]

The major limitation of the porcine preparations has been their large size and therefore their cost. This coupled with the lack of availability of gene-targeted models has meant that feeding of a high-fat diet for a very prolonged period has had to be performed.[38] Recently, a Danish group created a genetic porcine model of atherosclerosis by liver-specific overexpression of the D374Y gain-of-function mutant of the gene human PCSK9 in minipigs.[39] These animals develop hypercholesterolemia and human progressive lesions on a high-fat diet.

The PCSK9 protein binds to liver LDL receptors and targets them for degradation in lysosomes thus regulating LDL levels in the peripheral circulation. The mutation called D374Y (a gain-of-function mutation) increases the affinity of the binding of PCSK9 to LDLRs causing an LDL clearance defect leading to severe hypercholesterolemia in mice and humans.[40,41] The work of Bentzon *et al.*[39] reproduced this in pigs and this led to increased plasma cholesterol levels on regular chow which were further enhanced two- and threefold on a high-fat diet. The investigators assessed atherosclerosis in the aorta, left anterior descending coronary artery, and the iliofemoral artery according to the Virmani criteria for human plaques. This preparation appeared to recapitulate the pathophysiological features of human plaques such as those seen in humans, for example, preexisting intimal masses and intraplaque hemorrhage. The similarities between a human atherosclerotic coronary lesion and the same in the porcine preparation can be seen in Fig. 1.3.

Although an established drug target like a statin was not tested in this work, this is a major step forward for the field of atherosclerosis and this preparation should be useful for translational atherosclerosis research involving imaging, devices (such as stents), and where a human-like pathogenesis is essential.

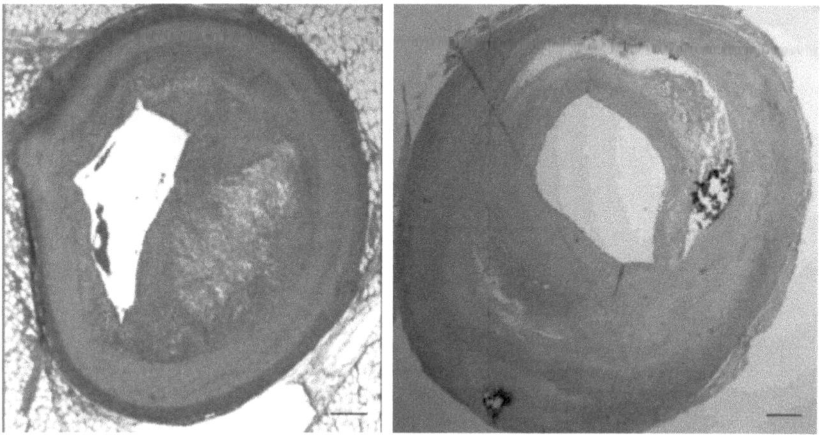

**Figure 1.3** Histological cross sections from left anterior descending coronary arteries after Elastic van Gieson staining. Panel (A) from a porcine experimental preparation and Panel (B) from a patient with human ischemic heart disease. Note the eccentric and similar extent and complexity of the thickening in both samples. *Panel A is with kind permission of Jacob Bentzon, Aarhus University, Denmark.* (See the color plate.)

## 10. WHAT HAVE WE LEARNED ABOUT THE BIOLOGY OF ATHEROSCLEROSIS FROM GENETIC EXPERIMENTAL PREPARATIONS SO FAR?

There are over 100 different gene-targeted mouse preparations with positive or negative effects on atherosclerosis ($LDLR^{-/-}$ or $APOE^{-/-}$), and these have been very effectively summarized in a recent review by Rader and colleagues.[42] Many more gene-targeted mouse preparations than this have been studied but those with no or nonstatistically significant effects on atherosclerosis are often not published in low-impact journals or not published at all. This leads to a strong publication bias and may lead to missed opportunities in atherosclerosis research. No one in the field has created an open-source repository for atherosclerosis data to date although the Mouse Phenome repository http://phenome.jax.org/ contains some information on the response of C57BL/6 strains to a high-fat, high-cholate (Paigen) diet.

In general, through the use of genetic preparations, the field now understands that the best-documented sufficient cause of atherosclerosis is hypercholesterolemia although additional factors (of which there are many) play key and necessary roles. We understand that the earliest lesion of atherosclerosis is the fatty streak, and through mouse preparations, we understand that the induction of adhesion molecules such as the selectins[43] and VCAM[44] plays a key role in this early stage. Chemotactic factors such as monocyte chemoattractant protein-1 induced by lipid products are induced soon thereafter. The consequence is an accumulation of monocytes (later T cells) inside arterial walls. Phenotypic modification of the monocytes occurs and these take up excess cholesterol esters and modified LDLs via scavenger receptors to become foam cells. This alters the balance of macrophages in the plaque such that their phenotype and number influence plaque presentation.[45]

## 11. HAS THE INFORMATION LEARNED ABOUT THE BIOLOGY OF ATHEROSCLEROSIS FROM MOUSE PREPARATIONS BEEN TRANSLATED TO THE CLINIC?

The information gleaned from the genetic preparations in mice has led to the conclusion that there are 30–40 proteins which impact upon atherosclerosis by alteration of lipid levels and probably two to three times as many that influence other pathways but not lipid levels.[42] There are still genes with unknown function so the numbers in both categories could yet rise further.

The translation of the information to the clinic has been relatively slow but there have been some successes within broad categories, for example, the statins, hypotensive drugs, drugs targeting nuclear hormone receptors, and anti-inflammatory approaches.[46] Interestingly, the genetic preparations do not always reveal the true effect in man. For example, despite >15 individual published mouse studies with five different statins, the overall broad conclusion from these is that there are no significant effects on lowering plasma cholesterol in the ApoE$^{-/-}$ system with a variable effect on atherosclerotic lesions depending on the statin used. Specifically, administration of pravastatin, fluvastatin, pitavastatin, cerivastatin, or simvastatin in the APOE$^{-/-}$ preparation had no effect on plasma cholesterol[47–49] but did alter lesion formation. Only the more potent statins such as atorvastatin had an effect on serum lipids on a background of a low-cholesterol experimental diet.[50] In the LDLR$^{-/-}$ preparation, statins also had variable and non-consistent effects on plasma cholesterol and atherosclerosis with some opposing data. Eventually, after a significant body of research, the field resolved that other pleiotropic effects of the statins on the liver and vessel walls were able to plausibly explain the beneficial effects.[51]

Similarly, the use of ACE inhibitors is widespread and effective in man[52] but data on atherosclerotic lesion size in the LDLR$^{-/-}$ and

$APOE^{-/-}$ preparations have been variant, with atherosclerosis reduced in $APOE^{-/-}$[53,54] but with no effect in the very few $LDLR^{-/-}$ preparations studied.[55] In case of the PPAR alpha agonists, data from experimental preparations with fenofibrate were largely positive,[56,57] and in man, this drug was weakly positive in selected groups of patients.[58]

Several other possible explanations for the discrepancies in the study outcomes between animals and human exist, for example, variations in the phenotypes of the patients in the clinical studies versus less variable phenotypes in inbred and near genetically identical animals. In addition, the time taken for the remodeling process to occur in humans is substantially longer than in genetically modified animals with a substantially shorter lifespan overall.

There is a recent example of an anti-inflammatory approach that has produced consistent data in mouse preparations, in large animal vascular injury and this is currently being evaluated in man. The interleukin-1 (IL-1) system of cytokines link with a number of fundamental cellular processes and the system is primarily controlled via the IL-1 receptor system. Work from our own laboratory showed that when the IL-1 receptor is deleted in an $APOE^{-/-}$ atherosclerosis preparation, although there are no effects on cholesterol, atherosclerosis is reduced by 40%.[59] The importance of the system was further exemplified in a porcine system where the naturally occurring IL-1 receptor blocker was administered and the response to angioplasty and stenting studied.[60] This has led to two clinical trials, one with the IL-1 receptor blocker, Kineret® called MRC IL-A HEART,[61] and a large trial CANTOS (Canakinumab anti-inflammatory thrombosis outcomes) using the first ever biologic in cardiovascular disease targeting one of the isoforms, IL-1beta, that binds to the IL-1 receptor.[62] Early data from IL-A HEART suggest a positive outcome upon the study endpoint, reduced area under the curve C-reactive protein at 14 days after non-ST elevation myocardial infarction with a potential rebound effect due to ongoing inflammation after the withdrawal of the therapy (unpublished).

There are some differences between atherosclerosis as studied in the mouse preparations and human atherosclerosis. A key difference is the site of lesion development (aorta in the mouse, coronaries and aorta in human), the mismatch between severity of the hypercholesterolemia in targeted mouse preparations and man, the disease time plan—progressive in mice but relapsing and remitting in humans over many years. In addition, the mouse preparations do not recapitulate occlusive disease with a transmural rupture endpoint leading to presentation as myocardial infarction.

Gene differences in mouse and man may also complicate matters when extrapolating from mouse to man. The cholesteryl ester transfer protein (CETP) which is responsible for transferring cholesteryl esters from HDLs to VLDLs and other proatherogenic particles is not expressed in mice; this makes all mouse systems antiatherogenic and inhibits progress to understand the biology behind the development of translational approaches to inhibit CETP in man.[63]

Another key difference having an impact on understanding from the GWAS is the lack of similarity between some regions of the mouse and human chromosomes. This is most starkly illustrated with the most replicated locus for early-onset myocardial infarction on human chromosome 9p21[64] where the SNP association is within the intergenic region. This region encodes the long antisense noncoding RNA in the INK4 locus (ANRIL) and the molecular mechanisms may be linked to Alu repeats as possible signposts of epigenetic modification.[65] The orthologous chromosome in mice is chromosome 4 but this is incompletely sequenced due to repetitive DNA and the candidate gene(s) in the region, ANRIL, have not yet been identified. Despite this, one study has deleted a 70 kb region of mouse chromosome 4 (assuming the gene of interest is there) and has shown that this has consequences upon vascular cell responses to inflammatory stimuli.[66]

## 12. FUTURE PERSPECTIVES

There is no doubt that over the last 20 years the cardiovascular research field has made great strides forward with understanding of the biology of atherosclerosis from using rodent preparations. Indeed, added to these findings, we have inevitably made a detailed study of the biology of the mouse.

However, given the data emerging from human GWAS, which indicate that nonlipid candidate genes are emerging equally alongside lipid genes, I believe it is time to reevaluate the rodent preparations to try and streamline their use. Three important issues that arise with these preparations are the use of extreme hypercholesterolemia (even in the milder $LDLR^{-/-}$ preparations), the lack of homology between mouse and man for key genes of interest, and the differences that have emerged between man and mouse with regard to immune processes.

It is time that researchers in the field start to think about new ways of developing atherosclerosis science by using a more human-translational approach. One suggestion might be that the research community develops and shares a digital representation of the atherosclerosis developed in mouse arteries (an avatar) such that multiscale processes (tissue, cell, and molecules) in artery walls can be represented, modeled, and linked with physiological data such as blood pressure and activity. This might obviate the need for so many rodent preparations and allow the field to work on other systems in depth, for example, atherosclerotic minipigs that are more akin to humans. There is still debate as to the importance of individual anatomy and therefore physiology to individual risk of atherosclerosis; henceforth, preparations that can replicate human anatomy and refocus here may allow these complex questions to be tackled using engineering tools.

Overall, while the various preparations of atherosclerosis have provided a number of useful clues and routes to treatments, the number of useful treatments of atherosclerosis that limit adverse consequences and disease presentation in man is limited, suggesting that there are causal pathways still undiscovered.

## ACKNOWLEDGMENTS

Thanks to Laura West from the Department of Cardiovascular Science at the University of Sheffield for unpublished images from mouse preparations of atherosclerosis and to Professor Jacob Bentzon from the Department of Cardiology, Aarhus University Hospital and Aarhus University, Denmark for unpublished porcine histological micrographs from D374Y-PCSK9 transgenic pigs.

## REFERENCES

1. Stone GW, Maehara A, Lansky AJ, et al. A prospective natural-history study of coronary atherosclerosis. *N Engl J Med.* 2011;364(3):226–235.
2. Samani NJ, Erdmann J, Hall AS, et al. Genomewide association analysis of coronary artery disease. *N Engl J Med.* 2007;357(5):443–453.
3. Schunkert H, Konig IR, Kathiresan S, et al. Large-scale association analysis identifies 13 new susceptibility loci for coronary artery disease. *Nat Genet.* 2011;43(4):333–338.
4. Davies MJ. Stability and instability: two faces of coronary atherosclerosis. The Paul Dudley White Lecture 1995. *Circulation.* 1996;94(8):2013–2020.
5. Virmani R, Kolodgie FD, Burke AP, Farb A, Schwartz SM. Lessons from sudden coronary death: a comprehensive morphological classification scheme for atherosclerotic lesions. *Arterioscler Thromb Vasc Biol.* 2000;20(5):1262–1275.
6. Ryding AD, Sharp MG, Mullins JJ. Conditional transgenic technologies. *J Endocrinol.* 2001;171(1):1–14.
7. Plump AS, Smith JD, Hayek T, et al. Severe hypercholesterolemia and atherosclerosis in apolipoprotein E-deficient mice created by homologous recombination in ES cells. *Cell.* 1992;71(2):343–353.
8. Piedrahita JA, Zhang SH, Hagaman JR, Oliver PM, Maeda N. Generation of mice carrying a mutant apolipoprotein E gene inactivated by gene targeting in embryonic stem cells. *Proc Natl Acad Sci USA.* 1992;89(10):4471–4475.

9. Ishibashi S, Brown MS, Goldstein JL, Gerard RD, Hammer RE, Herz J. Hypercholesterolemia in low density lipoprotein receptor knockout mice and its reversal by adenovirus-mediated gene delivery. *J Clin Invest.* 1993;92(2):883–893.
10. Powell-Braxton L, Veniant M, Latvala RD, et al. A mouse model of human familial hypercholesterolemia: markedly elevated low density lipoprotein cholesterol levels and severe atherosclerosis on a low-fat chow diet. *Nat Med.* 1998;4(8):934–938.
11. Ishibashi S, Herz J, Maeda N, Goldstein JL, Brown MS. The two-receptor model of lipoprotein clearance: tests of the hypothesis in "knockout" mice lacking the low density lipoprotein receptor, apolipoprotein E, or both proteins. *Proc Natl Acad Sci USA.* 1994;91(10):4431–4435.
12. Getz GS, Reardon CA. Diet and murine atherosclerosis. *Arterioscler Thromb Vasc Biol.* 2006;26(2):242–249.
13. Vergnes L, Phan J, Strauss M, Tafuri S, Reue K. Cholesterol and cholate components of an atherogenic diet induce distinct stages of hepatic inflammatory gene expression. *J Biol Chem.* 2003;278(44):42774–42784.
14. Klinge CM. Estrogen receptor interaction with estrogen response elements. *Nucleic Acids Res.* 2001;29(14):2905–2919.
15. Bourassa PA, Milos PM, Gaynor BJ, Breslow JL, Aiello RJ. Estrogen reduces atherosclerotic lesion development in apolipoprotein E-deficient mice. *Proc Natl Acad Sci USA.* 1996;93(19):10022–10027.
16. Daugherty A. Mouse models of atherosclerosis. *Am J Med Sci.* 2002;323(1):3–10.
17. Teupser D, Persky AD, Breslow JL. Induction of atherosclerosis by low-fat, semisynthetic diets in LDL receptor-deficient C57BL/6J and FVB/NJ mice: comparison of lesions of the aortic root, brachiocephalic artery, and whole aorta (en face measurement). *Arterioscler Thromb Vasc Biol.* 2003;23(10):1907–1913.
18. Zhang SH, Reddick RL, Piedrahita JA, Maeda N. Spontaneous hypercholesterolemia and arterial lesions in mice lacking apolipoprotein E. *Science.* 1992;258(5081):468–471.
19. van Ree JH, van den Broek WJ, Dahlmans VE, et al. Diet-induced hypercholesterolemia and atherosclerosis in heterozygous apolipoprotein E-deficient mice. *Atherosclerosis.* 1994;111(1):25–37.
20. Krieger M, Herz J. Structures and functions of multiligand lipoprotein receptors: macrophage scavenger receptors and LDL receptor-related protein (LRP). *Annu Rev Biochem.* 1994;63:601–637.
21. Nakashima Y, Plump AS, Raines EW, Breslow JL, Ross R. ApoE-deficient mice develop lesions of all phases of atherosclerosis throughout the arterial tree. *Arterioscler Thromb.* 1994;14(1):133–140.
22. Rosenfeld ME, Polinsky P, Virmani R, Kauser K, Rubanyi G, Schwartz SM. Advanced atherosclerotic lesions in the innominate artery of the ApoE knockout mouse. *Arterioscler Thromb Vasc Biol.* 2000;20(12):2587–2592.
23. McGillicuddy CJ, Carrier MJ, Weinberg PD. Distribution of lipid deposits around aortic branches of mice lacking LDL receptors and apolipoprotein E. *Arterioscler Thromb Vasc Biol.* 2001;21(7):1220–1225.
24. Brown MS, Goldstein JL. Lipoprotein receptors: therapeutic implications. *J Hypertens Suppl.* 1990;8(1):S33–S35, discussion S35–36.
25. Beisiegel U, Weber W, Ihrke G, Herz J, Stanley KK. The LDL-receptor-related protein, LRP, is an apolipoprotein E-binding protein. *Nature.* 1989;341(6238):162–164.
26. Teng B, Burant CF, Davidson NO. Molecular cloning of an apolipoprotein B messenger RNA editing protein. *Science.* 1993;260(5115):1816–1819.
27. Aparicio-Vergara M, Shiri-Sverdlov R, Koonen DP, Hofker MH. Bone marrow transplantation as an established approach for understanding the role of macrophages in atherosclerosis and the metabolic syndrome. *Curr Opin Lipidol.* 2012;23(2):111–121.
28. Burger PC, Wagner DD. Platelet P-selectin facilitates atherosclerotic lesion development. *Blood.* 2003;101(7):2661–2666.

29. Paigen B, Morrow A, Holmes PA, Mitchell D, Williams RA. Quantitative assessment of atherosclerotic lesions in mice. *Atherosclerosis*. 1987;68(3):231–240.
30. Bond AR, Jackson CL. The fat-fed apolipoprotein E knockout mouse brachiocephalic artery in the study of atherosclerotic plaque rupture. *J Biomed Biotechnol*. 2011;2011:379069.
31. Choudhury RP, Fayad ZA, Aguinaldo JG, et al. Serial, noninvasive, in vivo magnetic resonance microscopy detects the development of atherosclerosis in apolipoprotein E-deficient mice and its progression by arterial wall remodeling. *J Magn Reson Imaging*. 2003;17(2):184–189.
32. Trogan E, Fayad ZA, Itskovich VV, et al. Serial studies of mouse atherosclerosis by in vivo magnetic resonance imaging detect lesion regression after correction of dyslipidemia. *Arterioscler Thromb Vasc Biol*. 2004;24(9):1714–1719.
33. Bigalke B, Phinikaridou A, Andia ME, et al. PET/CT and MR imaging in a murine model of progressive atherosclerosis using 64Cu-labeled glycoprotein VI-Fc. *Circ Cardiovasc Imaging*. 2013;6:957–964.
34. Fang L, Green SR, Baek JS, et al. In vivo visualization and attenuation of oxidized lipid accumulation in hypercholesterolemic zebrafish. *J Clin Invest*. 2011;121(12): 4861–4869.
35. Gopalakrishnan K, Kumarasamy S, Abdul-Majeed S, et al. Targeted disruption of Adamts16 gene in a rat genetic model of hypertension. *Proc Natl Acad Sci USA*. 2012;109(50):20555–20559.
36. Granada JF, Kaluza GL, Wilensky RL, Biedermann BC, Schwartz RS, Falk E. Porcine models of coronary atherosclerosis and vulnerable plaque for imaging and interventional research. *EuroIntervention*. 2009;5(1):140–148.
37. Schwartz RS, Edelman E, Virmani R, et al. Drug-eluting stents in preclinical studies: updated consensus recommendations for preclinical evaluation. *Circ Cardiovasc Interv*. 2008;1(2):143–153.
38. Jensen TW, Mazur MJ, Pettigew JE, Perez-Mendoza VG, Zachary J, Schook LB. A cloned pig model for examining atherosclerosis induced by high fat, high cholesterol diets. *Anim Biotechnol*. 2010;21(3):179–187.
39. Al-Mashhadi RH, Sorensen CB, Kragh PM, et al. Familial hypercholesterolemia and atherosclerosis in cloned minipigs created by DNA transposition of a human PCSK9 gain-of-function mutant. *Sci Transl Med*. 2013;5(166):166ra1.
40. Kwon HJ, Lagace TA, McNutt MC, Horton JD, Deisenhofer J. Molecular basis for LDL receptor recognition by PCSK9. *Proc Natl Acad Sci USA*. 2008;105(6):1820–1825.
41. Soutar AK. Unexpected roles for PCSK9 in lipid metabolism. *Curr Opin Lipidol*. 2011;22(3):192–196.
42. Stylianou IM, Bauer RC, Reilly MP, Rader DJ. Genetic basis of atherosclerosis: insights from mice and humans. *Circ Res*. 2012;110(2):337–355.
43. Bourdillon MC, Randon J, Barek L, et al. Reduced atherosclerotic lesion size in P-selectin deficient apolipoprotein E-knockout mice fed a chow but not a fat diet. *J Biomed Biotechnol*. 2006;2006(2):49193.
44. Cybulsky MI, Iiyama K, Li H, et al. A major role for VCAM-1, but not ICAM-1, in early atherosclerosis. *J Clin Invest*. 2001;107(10):1255–1262.
45. Moore KJ, Sheedy FJ, Fisher EA. Macrophages in atherosclerosis: a dynamic balance. *Nat Rev Immunol*. 2013;13(10):709–721.
46. Zadelaar S, Kleemann R, Verschuren L, et al. Mouse models for atherosclerosis and pharmaceutical modifiers. *Arterioscler Thromb Vasc Biol*. 2007;27(8):1706–1721.
47. Sata M, Nishimatsu H, Osuga J, et al. Statins augment collateral growth in response to ischemia but they do not promote cancer and atherosclerosis. *Hypertension*. 2004;43(6):1214–1220.
48. Li Z, Iwai M, Wu L, et al. Fluvastatin enhances the inhibitory effects of a selective AT1 receptor blocker, valsartan, on atherosclerosis. *Hypertension*. 2004;44(5):758–763.

49. Bea F, Blessing E, Bennett B, Levitz M, Wallace EP, Rosenfeld ME. Simvastatin promotes atherosclerotic plaque stability in apoE-deficient mice independently of lipid lowering. *Arterioscler Thromb Vasc Biol.* 2002;22(11):1832–1837.
50. van De Poll SW, Romer TJ, Volger OL, et al. Raman spectroscopic evaluation of the effects of diet and lipid-lowering therapy on atherosclerotic plaque development in mice. *Arterioscler Thromb Vasc Biol.* 2001;21(10):1630–1635.
51. Liao JK, Laufs U. Pleiotropic effects of statins. *Annu Rev Pharmacol Toxicol.* 2005;45:89–118.
52. Dagenais GR, Pogue J, Fox K, Simoons ML, Yusuf S. Angiotensin-converting-enzyme inhibitors in stable vascular disease without left ventricular systolic dysfunction or heart failure: a combined analysis of three trials. *Lancet.* 2006;368(9535):581–588.
53. Hayek T, Attias J, Smith J, Breslow JL, Keidar S. Antiatherosclerotic and antioxidative effects of captopril in apolipoprotein E-deficient mice. *J Cardiovasc Pharmacol.* 1998;31(4):540–544.
54. da Cunha V, Tham DM, Martin-McNulty B, et al. Enalapril attenuates angiotensin II-induced atherosclerosis and vascular inflammation. *Atherosclerosis.* 2005;178(1):9–17.
55. Sharabi Y, Grossman E, Sherer Y, et al. The effect of renin-angiotensin axis inhibition on early atherogenesis in LDL-receptor-deficient mice. *Pathobiology.* 2000;68(6):270–274.
56. Duez H, Chao YS, Hernandez M, et al. Reduction of atherosclerosis by the peroxisome proliferator-activated receptor alpha agonist fenofibrate in mice. *J Biol Chem.* 2002;277(50):48051–48057.
57. Srivastava RA, Jahagirdar R, Azhar S, Sharma S, Bisgaier CL. Peroxisome proliferator-activated receptor-alpha selective ligand reduces adiposity, improves insulin sensitivity and inhibits atherosclerosis in LDL receptor-deficient mice. *Mol Cell Biochem.* 2006;285(1–2):35–50.
58. Ginsberg HN. The ACCORD (action to control cardiovascular risk in diabetes) lipid trial: what we learn from subgroup analyses. *Diabetes Care.* 2011;34(suppl. 2):S107–S108.
59. Chamberlain J, Francis S, Brookes Z, et al. Interleukin-1 regulates multiple atherogenic mechanisms in response to fat feeding. *PLoS One.* 2009;4(4):e5073.
60. Morton AC, Arnold ND, Gunn J, et al. Interleukin-1 receptor antagonist alters the response to vessel wall injury in a porcine coronary artery model. *Cardiovasc Res.* 2005;68(3):493–501.
61. Crossman DC, Morton AC, Gunn JP, et al. Investigation of the effect of Interleukin-1 receptor antagonist (IL-1ra) on markers of inflammation in non-ST elevation acute coronary syndromes (The MRC-ILA-HEART Study). *Trials.* 2008;9:8.
62. Ridker PM, Thuren T, Zalewski A, Libby P. Interleukin-1beta inhibition and the prevention of recurrent cardiovascular events: rationale and design of the Canakinumab Anti-inflammatory Thrombosis Outcomes Study (CANTOS). *Am Heart J.* 2011;162(4):597–605.
63. Chapman MJ, Le Goff W, Guerin M, Kontush A. Cholesteryl ester transfer protein: at the heart of the action of lipid-modulating therapy with statins, fibrates, niacin, and cholesteryl ester transfer protein inhibitors. *Eur Heart J.* 2010;31(2):149–164.
64. Kathiresan S, Voight BF, Purcell S, et al. Genome-wide association of early-onset myocardial infarction with single nucleotide polymorphisms and copy number variants. *Nat Genet.* 2009;41(3):334–341.
65. Holdt LM, Hoffmann S, Sass K, et al. Alu elements in ANRIL non-coding RNA at chromosome 9p21 modulate atherogenic cell functions through trans-regulation of gene networks. *PLoS Genet.* 2013;9(7):e1003588.
66. Harismendy O, Notani D, Song X, et al. 9p21 DNA variants associated with coronary artery disease impair interferon-gamma signalling response. *Nature.* 2011;470(7333):264–268.

CHAPTER TWO

# Genetics of Cardiovascular Development

**Bill Chaudhry, Simon Ramsbottom, Deborah J. Henderson**
Institute of Genetic Medicine, Newcastle University, Newcastle, United Kingdom

## Contents



## Abstract

Structural malformations of the heart are the commonest abnormalities found at the time of birth and the incidence is higher in fetuses that are lost during the first trimester. Although the form of the heart has been studied for centuries, it is in the past decades that the genetic pathways that control heart development have been unraveled. Recently, the concept of the second heart field, a population of multipotent cardiac cells that augment the initial simple heart tube, has clarified the development of the heart. Understanding how the second heart field is used in morphogenesis and how genes interact in a subtle and more complex way is moving us closer to understanding how the normal heart forms and why abnormalities occur. In this chapter, we present a description of the morphological processes that create the formed postnatal human heart and emphasize key genetic pathways and genes that control these aspects. Where possible, these are also linked to the common patterns of human cardiac malformation. Undoubtedly, the details will refine or change with further research but emphasis has been placed on areas of greatest certainty and the presentation designed to promote a general understanding.

*Progress in Molecular Biology and Translational Science*, Volume 124
ISSN 1877-1173
http://dx.doi.org/10.1016/B978-0-12-386930-2.00002-1

## 1. EARLY MORPHOGENESIS OF THE MAMMALIAN HEART

The cells that initially form the heart are of mesodermal origin and derive from common progenitors that also give rise to the hematopoietic lineages. The genetic pathways that orchestrate these initial steps are well known. In vertebrates, the *Eomesodermin* (*Eomes*) transcription factor is expressed before gastrulation (formation of the three embryonic layers—ectoderm, mesoderm, and endoderm) and is the earliest factor that has been shown to be required for cardiac induction.[1] *Eomes* induces the expression of the transcription factor, *Mesp1*, which acts as a master regulator of cardiogenesis and results in the downregulation of stem cell-related pluripotency genes and the upregulation of cardiac-specific genes including *Nkx2.5* and *Gata5*. These two genes have been shown to be critical for the specification and migration of cardiac precursors prior to gastrulation.[2–4] Interactions with signals from the surrounding tissues, in particular endoderm, are also required to form cardiomyocytes (Fig. 2.1). Factors produced by the endoderm and implicated as inducing expression of myocardial-specific genes include bone morphogenetic proteins (BMPs) and fibroblast growth factors (FGFs); FGF4 and FGF8 are expressed in the anterior endoderm[5,6] and are capable of inducing the expression of myocardial-specific markers in combination with BMP2.[7,8] Indeed, synergy between FGF and BMP signaling appears to be essential for myocardial differentiation in the primary heart fields. Other pathways, and in particular canonical Wnt signaling, are thought to be inhibitory to myocardial differentiation. The role played by canonical Wnt signaling during early cardiogenesis is complex (see review in Ref. 9). It is essential for early mesoderm formation, but inhibitory for the induction of early myocardial markers. For example, in mouse models, blocking Wnt signaling by targeted deletion of the canonical Wnt downstream mediator beta-catenin results in multiple hearts being formed at the expense of endoderm.[10] In contrast, Wnt5a and Wnt11, non-canonical (β-catenin-independent) Wnt ligands, are required to induce cardiogenesis in the second heart field (SHF) in mouse embryos.[11] Thus, complex interactions of inductive and repressive factors are involved in cardiac induction. As cardiac precursors sweep through the lateral plate mesoderm to the midline, they transiently form the cardiac crescent. This was thought to represent the full complement of cardiomyocytes, but it was shown in 2001 that a population of cells that comprise what is now known

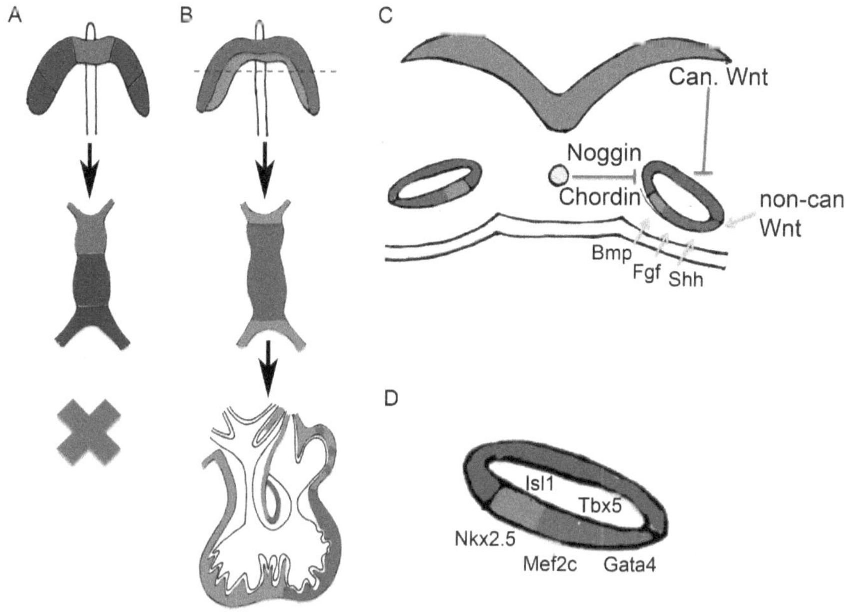


**Figure 2.1** Cardiac fields and cardiomyocyte induction. (A) Segmental model of heart formation. Until recently, it was thought that the cells within the cardiac crescent were prespecified to become particular regions of the heart. This model has now fallen out of favor. (B) Heart field model. It is now widely accepted that cardiac progenitors originate in two fields either side of the midline of the early embryo. The primary heart field cells (green) form the linear heart tube but are restricted to the left ventricle, the atrioventricular canal, and to some extent to the atria in the formed heart. This is because the secondary heart field cells (orange) are added to the poles of the heart tube and contribute to the outflow tract, right ventricle, and the majority of the atria in the formed heart. Dotted line indicates the position of the section in (C). (C) Secreted inhibitory factors (red block arrows) from the neural tube (pink) and notochord (cyan) block cardiac mesoderm formation, whereas positive signals (green arrows) from the endoderm (yellow) and lateral plate mesoderm induce cardiac mesoderm formation. (D) Key transcription factors induced in the cardiac mesoderm initiate a signaling cascade that results in the formation of functional cardiomyocytes. (See the color plate.)

as the SHF lie anteroinferiorly to the cardiac crescent (or primary heart field).[12–14] These cells do not form the primary heart tube but add on much later to the formed heart (see below and Fig. 2.1).

Less is known about the factors that induce cardiac progenitors to differentiate into endothelial cells.[15] In the early mouse embryo, precardiac cells in the bilateral cardiac fields give rise to both endocardial and myocardial progenitors.[16] Although endocardial progenitors are difficult to track in mouse embryos because of their diverse origins, it has been shown that both cardiomyocytes and endocardial cells can arise from a common Flk+ progenitor population.[17] The common origin of these lineages has been supported by lineage tracing using the SHF-specific lines *Isl1-Cre* and *Mef2c-Cre*, which have shown that both myocardial and endocardial cells are derivatives of these lineages.[18,19] The Ets-family member, *Ets2*, has been shown to be induced by the cardiac transcription factor *Nkx2.5* and be a key gene for endothelial–endocardial specification.[20] Similarly, the transcription factor *Scl/Tal1* has also been shown to be crucial for formation of the endocardium. These and other studies have confirmed that endocardial cells arise from an endothelial lineage by the process of vasculogenesis (i.e., *de novo* induction of endothelial cells rather than outgrowth from existing vessels).[20,21]

Endocardial cells and cardiomyocytes from the first heart field migrate to the midline to form a primitive heart tube inner composed of an inner endocardial layer and outer myocardial layer separated by a prominent layer of extracellular matrix known as cardiac jelly.[22] At these early stages, it is likely that the abundant cardiac jelly is important in promoting unidirectional peristaltic pumping. This simple but functional heart then undergoes progressive refashioning. Elongation of the heart tube, which is only tethered to the embryo at each end, is associated with bowing of the heart tube forward. This, with concomitant rightward rotation, is known as the process of looping.[23] While looping, specialized atrial and ventricular cardiac chamber myocardium begins to develop on the outer curvature of the looped heart tube; by the end of heart development, the initial heart tube will be represented by a small part of the atria, the atrioventricular canal, and the left ventricle (Fig. 2.1). The atrial and ventricular chambers emerge by a process of localized tissue expansion, which involves proliferation and shape changes. This process has been termed ballooning.[24] It is likely that abnormalities disturbing these early morphological processes are incompatible with further remodeling of the heart and are therefore not associated with survival beyond the first trimester. For example, *Nkx2.5* null mice are able to form a primitive heart tube, probably because there is functional redundancy from other Nkx family members, but development then arrests at this stage.[25,26]

These early transcriptional events have, however, been the topic of intensive research in the past decade as efforts have been made to differentiate cardiomyocytes from embryonic stem cells. Unexpectedly, only a few key transcriptional factors need to be expressed within a cell to drive differentiation toward the myocardial lineage. Recent studies have shown that a chromatin remodeling factor, *Baf60c*, together with the transcription factor, *Gata4*, is sufficient to induce the expression of cardiac markers from nonspecified mesoderm.[27] The addition of *Tbx5* is sufficient to induce beating. Together, these factors specifically induce the expression of *Nkx2.5*, one of the earliest and best understood markers of myogenic precursors in vertebrates. However, more recent studies have shown that myocardial transcription is directed by complexes containing multiple transcriptional regulators, including Nkx2.5, GATA4, Tbx5, SRF, and Mef2A, as well as Smarcd3, an epigenetic regulator of cardiac transcription, that bind to specific enhancers upstream of cardiac-specific genes to regulate localized differentiation of cardiomyocytes.[28] Thus, the absence of a single member may decrease transcriptional efficiency but does not block it entirely. This may provide some explanation of the variable penetrance seen in inherited cardiac malformations.

## 1.1. The second heart field

Until 15 years ago, the accepted view of heart development was that each segment of the heart (atria, left ventricle, right ventricle, and outflow tract) was already specified in the primitive heart tube. The identification of the SHF has changed this view. The heart starts as an inlet chamber and an outlet chamber and material is added on to each end, augmenting both poles of the heart (Fig. 2.1). Interestingly, the areas that are augmented are those that relate to the specialization of the heart to pump in a parallel circulation using lungs for oxygen exchange. Thus, the SHF-derived cells add to the atria, greatly to the right ventricle and to the outflow tract. The SHF is initially found adjacent to the cardiac crescent but the physical location of this region eventually changes to become identified as the pharyngeal mesoderm, as the cranial part of the embryo grows and folds. Once the cells have added to the poles of the heart, they must then differentiate to cardiomyocytes, endocardium, and smooth muscle cells. It is not yet known whether this late differentiation requires different signals compared to the primary heart field cells.

SHF cells share many features in common with the cells of the primary heart fields (or cardiac crescent)—indeed it has been argued that they form part of a continuous mass of cells and that the divisions based on overlapping gene expression discussed below are artificial.[29] However, the gene expression differences that allow the cells derived from the first and second heart fields to be distinguished hint at differences between the populations and perhaps discrete roles. Many signaling pathways are experienced both by primary heart field cells as well as SHF cells—although the latter group may have longer exposure to these transcription factors. These include Wnt, FGF, BMP, Sonic Hedgehog (Shh), Notch, and retinoic acid. Although there is now an extensive literature surrounding the SHF, and this has been reviewed extensively recently,[30–32] some key points will be discussed here. Both canonical and noncanonical Wnt signaling have been shown to be important within the SHF, with canonical Wnt signals playing essential roles in maintaining proliferation and preventing premature differentiation of SHF progenitors.[33–35] Noncanonical Wnt signals (Wnt5a and Wnt11) act to restrict the action of canonical Wnt and promote differentiation.[11] Other factors, such as FGFs, BMPs, retinoic acid, and Shh, also act to regulate the size of the SHF precursor population (see reviews in Refs. 30–32). Thus, a complex array of signaling factors is involved in maintaining expansion and the later differentiation of the SHF population. More recently, it has been shown that the SHF can also be considered to comprise anterior and posterior elements, with the anterior portion contributing mainly to the arterial pole of the heart and the posterior to the venous pole.[36] As one might expect, although the majority of genes overlap between the two populations, distinct patterns of gene expression are emerging (see review in Ref. 31). These patterns of expression are strongly affected by gene pathways that define the anterior–posterior identity of tissues. Two well-known examples are retinoic acid signaling and Hox gene identity. Excessive retinoic acid has been suggested to "posteriorize" the developing heart tube, manifesting as a loss of outflow components and expansion of the inflow components.[37,38] More recently, retinoic acid has been shown to regulate the expression of specific Hox genes within the SHF that define specific subdomains at both the outflow and inflow regions of the heart.[39]

Two genes, *Isl1* and *Tbx1*, are however considered characteristic of the SHF. The transcription factor *Isl1*, which is found only transiently in the first heart field, is expressed throughout the SHF.[18] This prolonged expression is in keeping with the greater expansion that is required from the SHF lineage.

With recognition of the SHF has come awareness that common human malformations in different parts of the heart might relate to abnormalities in this lineage. As there are no long-term persistent molecular signatures in SHF-derived tissues, much understanding has been derived from mouse models with disruption of SHF-restricted genes or key signaling factors implicated in regulating SHF cell maintenance and recruitment to the poles of the heart.[40] As expected, SHF defects can affect both the inflow and

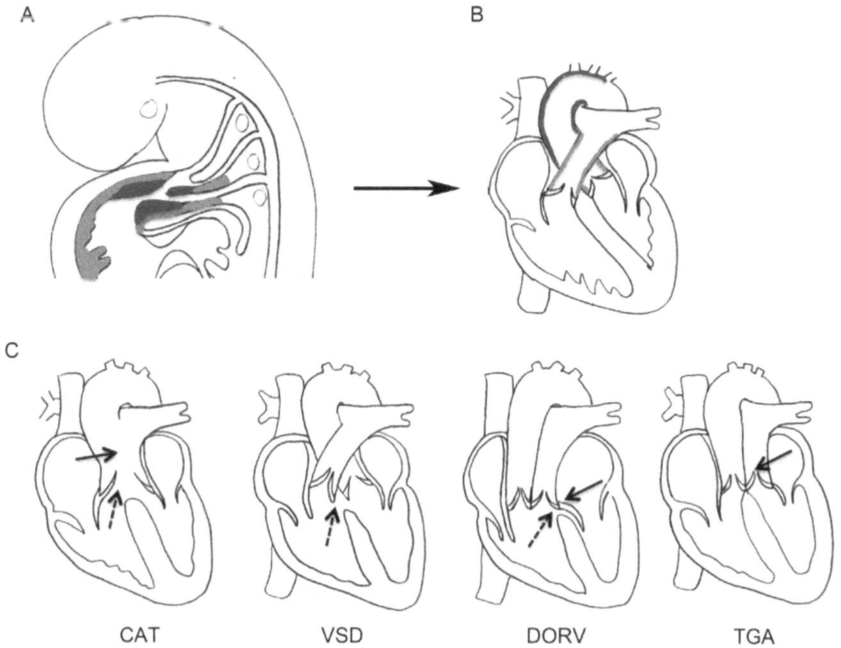


**Figure 2.2** Outflow tract septation and associated defects. (A) Cells from the second heart field (orange) form the outflow tract myocardium, whereas cells from the neural crest (purple) make a major contribution to the outflow cushions and the dorsal wall of the aortic sac. These neural crest-filled tissues fuse to initiate outflow tract septation. (B) In the formed heart, the smooth muscle layers of the ascending aorta and pulmonary trunk are mostly derived from the second heart field (orange), whereas the aortic arch smooth muscle is derived from neural crest cells (purple). (C) Abnormalities in both the second heart field and neural crest cell contributions give rise to an overlapping spectrum of outflow tract anomalies that affect different regions of the great arteries. In common arterial trunk (CAT), the outflow vessel fails to divide resulting in a single outflow vessel. In double outlet right ventricle (DORV), the outflow tract divides but the spiraling of the arterial trunks does not occur and both the aorta and pulmonary trunk communicate almost exclusively with the right ventricle. Both of these malformations are associated with ventricular septal defects (VSDs). In transposition of the great arteries (TGA), spiraling of the great arteries does not occur, but in this case, the pulmonary trunk is associated with the left ventricle, whereas the aorta communicates with the right ventricle. (See the color plate.)

outflow of the heart, either together or in isolation. At the outflow, typical defects include common arterial trunk, double outlet right ventricle, interrupted aortic arch, and ventricular septal defects (VSDs, Fig. 2.2); and at the venous pole, the most commonly associated defects are atrioventricular septal defects (AVSDs), arising from the loss of the dorsal mesocardial protrusion

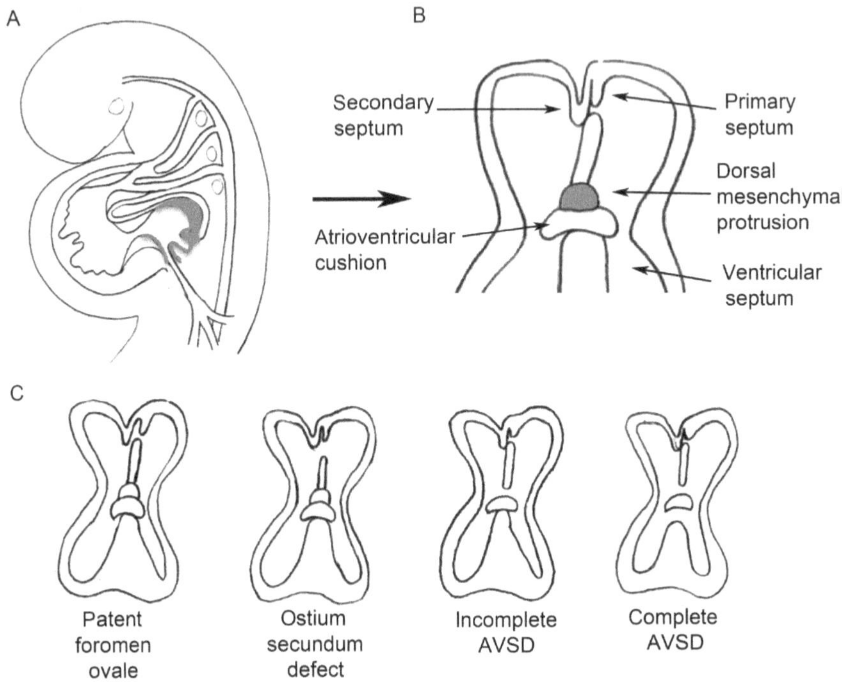


**Figure 2.3** Atrial septation. (A) Cells from the second heart field (orange) migrate into the venous pole of the heart and make a major contribution to atrial septation. (B) Structures involved in atrial septation. (C) Disruption of the components of atrial septation results in abnormalities of atrial or atrioventricular septation. In patent foramen ovale, the structures are present but the primary atrial septum does not adhere to the secondary septum. In ostium secundum defects, the primary atrial septum does not overlap with the secondary septum. In incomplete atrioventricular septal defects, the dorsal mesenchymal projection is deficient. In complete atrioventricular septal defects, fusion of the ventricular septum is also affected. (See the color plate.)

(DMP) rather than the primary or secondary atrial septae (Fig. 2.3). In all of these defects, the primary abnormality appears to result in a deficiency in the number of cells adding from the SHF and thus a hypoplasia in the tissues that are normally derived from them. These are discussed in more detail in each section below. Human mutations have been described in *ISL1*, although they do not lead to complete loss of ISL1 protein production.[41] This is perhaps not surprising as mouse knockouts indicate that loss of *Isl1* results in a small unlooped heart with major deficiencies in atrial and outflow components (outflow tract and right ventricle)[18]; this would not be compatible with intrauterine life. *TBX1* is a key gene associated with human malformation of the SHF. Disruption of its mouse homologue, *Tbx1*, causes cardiovascular malformations in keeping with DiGeorge syndrome, including common arterial trunk and interrupted aortic arch,[42–44] but surprisingly few specific mutations have been found, suggesting that the regulation of this gene by other interacting genes present within the DiGeorge syndrome deletion on chromosome 22q11 will turn out to be important in the etiology of the disease. The mechanisms by which *Tbx1* deficiency leads to cardiovascular malformations are being unraveled using knockout mice (see later).

## 2. DEVELOPMENT OF THE VENTRICULAR WALL

The ballooning ventricular wall progressively enlarges and thickens during development through addition from the SHF and also local cardiomyocyte proliferation. The original primary heart tube persists at the inner curvature but becomes comparatively smaller as development progresses. Many transcription factors, including *Nkx2.5*, *Tbx5*, *Gata4*, and *Tbx20*, are expressed in both the primary and chamber myocardium and have been shown to act together as cofactors to form multiprotein transcriptional complexes that regulate differentiation of the myocardium.[28,45] The gene encoding atrial natriuretic peptide (*Nppa*) is one of the earliest markers of the chamber myocardium and is not only induced by these factors in the ballooning areas but also actively repressed by *Tbx2* and *Tbx3* in the nonchamber primitive tube myocardium.[46–48]

The ventricular chambers are characteristically thick walled and composed of an outer compact layer of cardiomyocytes and an inner trabeculated layer (coarse trabeculations are found in the right ventricle). The development of the ventricular wall myocardium is dependent on the epicardium—a layer of epithelioid cells that adhere to the outer surface of the heart (see review in Ref. 49). These cells originate from the septum transversum—which will later form the diaphragm. A lobulated mass of cells, the proepicardial organ, appears at this location and cells detach and travel either by migration over the surface of the pericardial coelom or float as vesicles until they adhere to the surface of the heart. Once attached, they rapidly divide and spread. The cells of the epicardium change from their epithelial phenotype and exhibit mesenchymal behavior allowing them to separate from the epicardial monolayer and invade the myocardium. The epicardium gives rise to the fibroblasts that are found throughout the myocardium and also the smooth muscle cells[50,51] and at least some of the endocardium of the coronary circulation.[52] The layer of epicardium signals to the underlying myocardium causing proliferation and increasing layers of cells.

A number or transcriptional regulators have been shown to be essential for this process, of which Wilm's tumor gene, *Wt1*, is the best studied. *Wt1* prevents the premature differentiation of epicardial cells by regulating cell–cell adhesion and the levels of key signaling molecules and again, in its absence the ventricular wall remains thin.[53,54] The compact layer undergoes rapid cell division regulated largely by epicardial-derived signals. These include retinoic acid, FGFs, Wnts, and erythropoietin.[55–58] In the absence of these factors, the myocardial wall is hypoplastic and the heart generally fails in mid-late gestation.

Inside the ventricular chamber, trabeculae, which are fingers of myocardium, extend into the chamber rendering the primitive ventricle morphologically distinct from the smooth-walled atrium. At this stage, they may play a role in nutrient transfer by increasing surface area or act as preferential channels for conduction of the cardiac impulse. The trabeculae form by delamination and directional cell migration of cardiomyocytes from the ventricular wall.[59–61] Signaling between the endocardium and the myocardium is known to be essential for the formation of the trabeculae and the specialization of the ventricular chamber. Neuregulin and its receptor, erbB2, have been implicated in regulating trabeculae development as trabeculae fail to form in mice where erbB2 has been inactivated.[62–64] Similarly, Notch signaling (including both Notch1 and Notch2) has been shown to be essential for the activation of ventricular-specific genes and the formation of the trabeculae.[65] Recently, mutations in the Notch signaling component *MIB1* have been shown to cause left ventricular noncompaction, a cardiomyopathy that is characterized by an expanded trabecular layer with excessive intertrabecular recesses within the compact region of the ventricular wall.[66]

As the myocardium of the ventricle develops through trabeculation, some of these protruding up in the central region extend and become compacted together to form the muscular part of the interventricular septum to divide the common ventricular chamber into left and right parts. The location of this morphological separation is based upon molecular signals. *Tbx5* transcripts are found in the left ventricle.[67] The ventricular septum is formed at the boundary between the *Tbx5*-positive cells of the left ventricle and the *Tbx5*-negative cells of the right ventricle, suggesting it might play some role in the positioning of the septum. Indeed, when *Tbx5* is expressed ubiquitously throughout the right and left ventricles, then the ventricular septum does not form giving a single ventricle expressing left ventricle-specific markers.[68] Moving the boundary of expression of *Tbx5* into the presumptive right ventricle alters the position of the developing

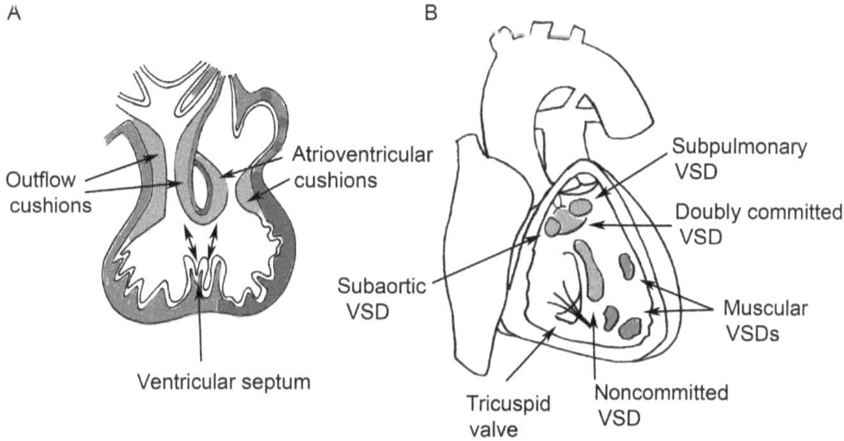


**Figure 2.4** Ventricular septation and ventricular septal defects. (A) Ventricular septation is achieved by the fusion of three structures: the muscular ventricular septum derived from the second heart field (orange) and the outflow and atrioventricular cushions (lilac). The myocardium derived from the primary heart field is indicated (green). (B) Ventricular septal defects can arise in either the muscular component of the septum (orange) or in the membranous part, derived from the cushions (lilac). (See the color plate.)

ventricular septum resulting in a small right ventricle and an expanded left ventricle whilst and creating a second boundary of *Tbx5* positive and negative cells in the right ventricle resulted in the formation of a second ectopic ventricular septum.[68] The boundary of *Tbx5* positive and negative cells does, therefore, appear to be important for the positioning of the ventricular septum and thus for ventricular specification. As discussed earlier, *Tbx5* interacts with other genes to bring about chamber-specific patterns of gene expression. For example, *Tbx5* acts synergistically with *Nkx2.5* and *GATA4* to activate the atrial natriuretic factor (*Nppa*) promoter.[68–70] Thus, the process of ventricular septation is intimately linked to chamber specialization and the main part of the septum is derived from ventricular myocardium. Abnormalities in the muscular part of the septum are known as muscular VSDs (Fig. 2.4).

## 3. ATRIAL DEVELOPMENT AND SEPTATION

Atrial septal defects are among the commonest cardiac defects and may be associated with devastating heart failure in the neonate or be remarkably silent until adulthood—recently being associated with migraine (see review in Ref. 71). SHF cells add to the venous pole of the primitive heart tube such that very little of the original inflow component of the linear heart tube is found in the mature atrial chambers (Fig. 2.3). The means by which the common atrium separates into two chambers is important to understand as there are several discrete processes, each involving different sources of embryonic tissue. This likely explains the failure to discover a single-gene pathway that causes atrial septal defects.

Separation of left and right atria is achieved by initial down-growth of a muscular septum (the primary atrial septum) from the roof of the atrium (Fig. 2.3). More posteriorly, the DMP, which is an SHF-derived structure that propagates from the back of the atrial wall,[72] fuses with the primary septum. Inferiorly, both these structures meet with local expansions of cardiac jelly at the atrioventricular junction, known as cushions, to separate the right and left atrial chambers. The temporary hole lying between the atrioventricular cushions, the DMP and the primary septum, is known as the ostium primum. As the primary atrial septum and DMP meet the atrioventricular cushions, closing the ostium primum, small perforations develop within the superior part of the primary septum. These coalesce to form a hole—the ostium secundum. On the right side of the primary septum, a marked infolding of the atrial roof (the secondary atrial septum) descends and partly covers the ostium secundum. This leaves an opening in the atrial septum, when viewed from within the right atrium, which is known as the foramen ovale. This persisting communication between the atria, leading to the ostium secundum, allows oxygenated and nutrient-rich blood returning to the fetus from the umbilical vein and inferior vena cava, to preferentially stream through these communications between the right and left atria, and out to the body. The inferior part of the primary septum overlaps with the secondary septum and is able to act as a flap valve[73] (Fig. 2.3). This is kept open by the positive pressure gradient prior to birth, but closes as left atrial pressures increase following birth.

With understanding of these processes, the spectrum of atrial septal defects can be explained. In some individuals, the septum secundum is mildly deficient, or the perforations within the superior part of the primary atrial septum are excessive, so that the flap valve cannot properly separate right and left atria. This is a patent foramen ovale and has been associated with migraine and paradoxical embolus. An ostium secundum defect is a larger and more direct communication between the atria—either caused by an abnormally large ostium secundum or a very deficient secondary septum. Failure of fusion between the primary septum, DMP, and cushions is known as an ostium primum defect (also known as an incomplete AVSD). When there is also a failure of ventricular septation, it is known as a complete AVSD (Fig. 2.3).

*TBX5* mutations are notably associated with atrial septal defects, and during early development, *Tbx5* is restricted to the most posterior part of the primitive heart tube that will give rise to the atria and sinus venosus. Dominant mutations in *NKX2.5* have been found in patients with atrial septal defect and VSDs, tetralogy of Fallot, and Ebstein's anomaly of the tricuspid valve.[74–76] Mutations in the *GATA4* gene have also been found in patients with atrial septal defects.[77] Interestingly, one of the reported mutations in *GATA4* disrupts interactions with the *TBX5* gene, suggesting interactions between these genes might be essential to bring about correct atrial septation.[77]

## 4. ATRIAL VENOUS CONNECTIONS

The atria initially attach to the venous circulation via bilateral superior and inferior caval veins entering at the lateral aspects of a common atrium. Growth and remodeling result in loss of the left-sided component, but the right venous system persists and becomes the superior and inferior caval veins connecting to the right atrium. Left carotid and subclavian veins lead to the left brachiocephalic vein, which drain via the right superior vena cava. The caval system draining the lower body persists as the hemiazygous system, which drains via the right interior vena cava. The remnant of the left superior caval vein that courses along the atrioventricular groove gives rise to the coronary sinus, which drains venous blood derived from the coronary circulation into the right atrium. The primitive pulmonary veins initially connect to the atria in the midline, while growth and remodeling lead to translocation of the pulmonary veins into their usual position in the posterior wall of what will be the left atrium (see review in Ref. 73).

### 4.1. The atrioventricular canal

The right and left ventricles are already separated to a great extent by the muscular interventricular septum arising from the apex of the heart. Through much of development there is a physiological communication between the ventricles at the tip of the septum allowing free mixing of right and left ventricular blood as it travels through a common outflow tract. As the outflow septates to give rise to separate aortic and pulmonary

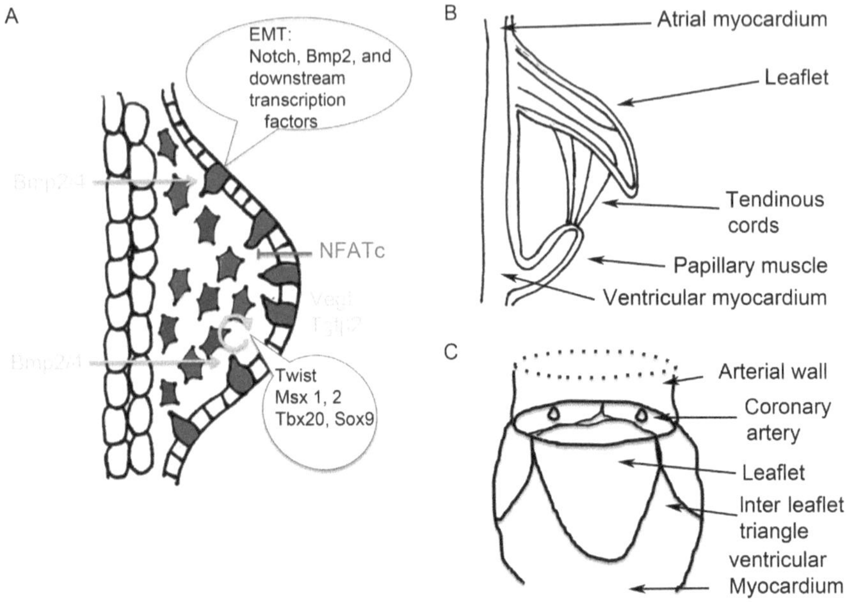


**Figure 2.5** Cushion formation and the cardiac valves. (A) Bmp signals from the myocardium induce the process of endothelial–mesenchymal transformation (EMT) in the overlying endocardium (in conjunction with TGFβ2 and VEGF from the endocardium) which results in the migration of cells from the endocardium into the cardiac jelly (blue). In contrast, NFATc signaling from the endocardium limits EMT. In the outflow cushions, neural crest cells (purple) are also present. Continued migration and proliferation of these cells forms the primitive valve structures. (B) Structure of a mature (generalized) atrioventricular valve leaflet showing its association with the myocardium. (C) Structure of the aortic valve showing the trifoliate nature of the leaflets and the associated coronary arteries. Below the valve is ventricular myocardium, whereas above it is arterial wall. (See the color plate.)

artery/ductus arteriosus flow, the ventricles complete septation. Both septation of the outflow and the ventricles is dependent on fusion of cardiac cushions at what was the inner curvature of the heart, an area known as the atrioventricular canal (Fig. 2.4). The factors that are required for formation of these cushions have been studied in detail (see comprehensively review in Ref. 78). Signals from the myocardium, including TGFβ family members, VEGF, and Notch, result in a transformation of cells in the adjacent myocardium to transform into mesenchyme and migrate into the cardiac jelly in between (Fig. 2.5). The deposition of an appropriate hyaluronan and proteoglycan-rich extracellular matrix has been shown to be essential for the normal development of the cushions as in the absence of these ECM proteins the cushions do not form.[79,80] Continued migration of cells from the endocardium, together with rapid proliferation, results in the formation of primitive valve-like structures that allow only unidirectional flow of blood through the heart. Ventricular septation itself is completed by the fusion of the interventricular muscular septum with the atrioventricular cushions and the proximal outflow cushions. The tissues derived from the cushions become fibrous tissue and are the membranous part of the interventricular septum in the formed heart. Deficiencies in this process give rise to subaortic, subpulmonary, and doubly committed membranous VSDs, depending on whether they sit under the aorta, pulmonary artery, or between both (Fig. 2.4). Given, the number of structures that must fuse in order to separate the atria from the ventricles, and each into left and right sides, it is not surprising that malformations in this area are relatively common. The most severe situation is an AVSD, where mixing of left and right atrial flows is complicated with mixing of ventricular flows and is very common in Down syndrome (see below).

The formation of the DMP has been shown to be dependent on both BMP and Shh signaling. In the absence of these important signaling molecules from the SHF precursors, the DMP does not form resulting in atrial septal defect and AVSD.[81,82] Disruption of a number of other genes specifically in the SHF results in similar defects,[72,83] highlighting the importance of the SHF for atrioventricular septation.

## 5. OUTFLOW TRACT SEPTATION AND THE NEURAL CREST

The outflow tract is formed as a single tube connected to the right ventricle. Cushions appear along the length of the vessel, formed in a similar way to that in the atrioventricular canal (Figs. 2.2 and 2.5). However, in addition to cells from the endocardium, the outflow tract cushions have a major contribution from another embryonic lineage (see review in Ref. 84). Neural crest cells (NCCs) are a pluripotent mesenchymal cell type that transforms from the neuroectoderm of the neural tube. These cells are capable of long-range migration and form a myriad of different cell types ranging from autonomic nerves in the distal gut to melanocytes that pigment skin and hair, as well as most of the structures of the head. A specific population of NCCs originating from the cranial end of the neural tube migrates through the pharyngeal arches into the outflow cushions where they are essential for bulking out the outflow cushions and the initiation of septation.[85] The initial fusion event that initiates outflow tract septation occurs between the NCC-packed area of the outflow tract cushions and the dorsal wall of the aortic sac.[86] This is followed by rapid progressive fusion of the proximal outflow tract cushion finishing with the fusion to the interventricular septum and atrioventricular cushion complex. Most of the NCCs that enter the outflow tract are destined to die by apoptosis once septation is complete.[87] As they disappear, their place is taken by myocardium from the outflow tract wall that extends inward to completely separate the proximal part of the outflow into two muscular tubes. This process of myocardialization continues until a separate infundibulum of the pulmonary trunk and an outlet of the left ventricle are created.[88] A minority of NCCs contribute to the arterial valves and the proximal coronary arteries.[89]

All of these septation processes are dependent on sufficient NCCs entering the outflow and mouse models of NCC insufficiency demonstrate a common arterial trunk; clinically this appearance is known as truncus arteriosus. Similarly, mouse models also show that there needs to be sufficient length to the outflow tract (i.e., sufficient SHF addition) to ensure that the outflow vessels can properly connect with the ventricles. Failure of normal outflow lengthening may be the cause of double outflow right ventricle, where both aorta and pulmonary artery arise from the right ventricular outflow (Fig. 2.2).

## 6. THE CONDUCTION SYSTEM

During development, the heart changes from a linear tube with peristaltic action to a multichambered three-dimensional organ functioning as a displacement pump. The direction of the depolarization wave causing contraction changes during development to reflect this, ensuring that ventricular contraction initiates at the apex. The propagation of the depolarization wave through the heart occurs through specialized conduction tissues, which in fact are the remnants of the primitive heart tube, which have not differentiated to chamber myocardium (see review in Ref. 90). This is the source of cells that differentiate to form the atrioventricular node, the atrioventricular bundle and its branches; the principle central components of the cardiac conduction system.[91] More peripherally, it is the trabeculations that act as preferential channels for depolarization, while the area in the heart with the fastest native rate of depolarization, the pacemaker or sinoatrial node, has been shown to originate from the SHF. The complex transcriptional regulation involved in the development of the cardiac conduction system has benefited from intense investigation over recent years.[90] A detailed description is beyond the scope of this chapter, but involves a largely overlapping set of transcriptional regulators that are also involved in chamber specialization, with spatially and temporally regulated activation and repression of genes that ultimately lead to the expression of the cardiac sodium channel genes *Scn5a* and *Scn10a* that are known to be essential for cardiomyocyte excitability.[45,92]

## 7. CARDIOVASCULAR DEFECTS IN THE CONTEXT OF SYNDROMES

Developmental malformations of the heart commonly occur as part of a complex or syndrome. In some circumstances, a clear genetic link is apparent although the actual genes or mechanisms involved remain elusive. The patterns of malformation also suggest there might be specific processes that underlie the spectrum of malformations seen in each situation. For example, in Down syndrome (trisomy 21) as well as craniofacial dysmorphology, mental retardation, hypotonia, short metacarpals, and phalanges, there is a high incidence of congenital heart disease—commonly VSDs, but also atrial septal defects and patent arterial duct. However, there is a remarkably high incidence of atrioventricular septation defects. In Turner syndrome, there is loss of one X chromosome, in part or whole, and this is commonly mosaic throughout the individual's cells. These females have a very high incidence of malformations involving the aortic valve, hypoplasia of the left ventricle and aortic arch, and coarctation of the aorta. Noonan syndrome results from a mutation in *PTPN11* and the defects include pulmonary stenosis and conduction anomalies.[93] More intriguingly is the role played by modifier genes. In DiGeorge or velocardiofacial syndrome, the defects include persistent truncus arteriosus (common arterial trunk) and interrupted aortic arch, accompanied by aplasia or hypoplasia of the thymus, parathyroids, and/or thyroid glands with abnormal facies. Although these syndromes often occur in patients with microdeletions of chromosome 22q11, the gene from this region that appears most closely associated with abnormal cardiovascular development is *TBX1*. Mouse embryos with homozygous deletion of the *Tbx1* gene show a phenotype that resembles closely that seen in the DiGeorge syndrome.[42–44] Moreover, a bacterial artificial chromosome carrying the human *TBX1* gene was shown to partially rescue the conotruncal defects generated by a 1.5 Mb deletion of the region corresponding to 22q11 in mice. These data, together with the expression pattern of the *Tbx1* gene, in the pharyngeal arches and outflow tract of the heart, suggest a major role for *TBX1* in the molecular etiology of DiGeorge syndrome. Surprisingly, few individuals with mutation of *TBX1* have been identified.[94] It seems likely that other genes might be interacting with or modifying the activities of *TBX1*.

Mutations in the transcription factor *TBX5* have been identified in patients with Holt–Oram syndrome. This syndrome is characterized by defects of the upper limb, the conduction system, tetralogy of Fallot, and atrial septal defect and VSD. However, within a single family, this phenotype can be quite variable and some family members can manifest quite complex heart abnormalities such as hypoplastic left heart syndrome and tetralogy of Fallot.[95,96] This is potentially quite confusing as hypoplastic left heart syndrome results in a small left ventricle, aortic valve, and aortic arch, while tetralogy of Fallot affects the right ventricular outflow tract and includes a particular type of VSD. Both of these defects are, however, associated with abnormal positioning of the ventricular septum, supporting a role for TBX5 in this process. Direct evidence for a role for *Tbx5* in positioning of the ventricular septum has come from studies in mice, where misexpression of *Tbx5* altered the position of the ventricular septum and in some cases resulted in the formation of a second septum.[68]

## 8. CONCLUSIONS

Understanding the steps involved in heart development and the transcriptional regulation that controls these processes is an important step in designing strategies to screen and prevent congenital heart malformations. For example, the finding that many of the processes that are disturbed relate to cell proliferation may suggest a specific directed pharmacological approach. In addition to the primary purpose of this research, developmental studies are also impacting on adult disease. Knowledge of transcriptional pathways that make cardiomyocytes has also been used to differentiate stem cells to restore diseased myocardium. However, there are still more questions to answer about how to specifically create cardiomyocytes with ventricular, atrial, and conduction properties.

These types of investigations will be facilitated by the availability of new animal models with fully characterized genomes, for example, zebrafish (see Chapter 4). The relative simplicity of the heart in this species, together with ease of transgenic manipulation and induction of mutations should accelerate progress in understanding such fundamental questions. However, we will still need to look to mammalian models with septated heart and lungs, such as the mouse, to evaluate the potential for a human context. Finally, the rapidity and falling costs of genome-scale sequencing should make the translation of research findings into clinical diagnostic testing a realistic proposition within the next few years.

## ACKNOWLEDGMENTS

It is not possible to represent the work of all those who are moving this field forward in this limited account and we apologize to those who have not been mentioned. We would like to acknowledge our friend and colleague Professor Robert Anderson who has kept our molecular genetic approaches grounded in cardiac morphology.

## REFERENCES

1. Arnold SJ, Hofmann UK, Bikoff EK, Robertson EJ. Pivotal roles for eomesodermin during axis formation, epithelium-to-mesenchyme transition and endoderm specification in the mouse. *Development.* 2008;135(3):501–511.
2. Kitajima S, Takagi A, Inoue T, Saga Y. MesP1 and MesP2 are essential for the development of cardiac mesoderm. *Development.* 2000;127:3215–3226.
3. Bondue A, Lapouge G, Paulissen C, et al. Mesp1 acts as a master regulator of multipotent cardiovascular progenitor specification. *Cell Stem Cell.* 2008;3(1):69–84.
4. Costello I, Pimeisl IM, Dräger S, Bikoff EK, Robertson EJ, Arnold SJ. The T-box transcription factor Eomesodermin acts upstream of Mesp1 to specify cardiac mesoderm during mouse gastrulation. *Nat Cell Biol.* 2011;13(9):1084–1091.
5. Zhu X, Sasse J, McAllister D, Lough J. Evidence that fibroblast growth factors 1 and 4 participate in regulation of cardiogenesis. *Dev Dyn.* 1996;207:429–438.
6. Alsan BH, Schultheiss TM. Regulation of avian cardiogenesis by Fgf8 signaling. *Development.* 2002;129:1935–1943.
7. Lough J, Barron M, Brogley M, Sugi Y, Bolender DL, Zhu X. Combined BMP-2 and FGF-4, but neither factor alone, induces cardiogenesis in non-precardiac embryonic mesoderm. *Dev Biol.* 2000;178:198–202.
8. Barron M, Gao M, Lough J. Requirement for BMP and FGF signaling during cardiogenic induction in non-precardiac mesoderm is specific, transient, and cooperative. *Dev Dyn.* 2000;218:383–393.
9. Gessert S, Kühl M. The multiple phases and faces of wnt signaling during cardiac differentiation and development. *Circ Res.* 2010;107(2):186–199.
10. Lickert H, Kutsch S, Kanzier B, Tamai Y, Taketo M, Kemler R. Formation of multiple hearts in mice following deletion of b-catenin in the embryonic endoderm. *Dev Cell.* 2002;3:171–181.
11. Cohen ED, Miller MF, Wang Z, Moon RT, Morrisey EE. Wnt5a and Wnt11 are essential for second heart field progenitor development. *Development.* 2012;139(11):1931–1940.
12. Mjaatvedt CH, Nakaoka T, Moreno-Rodriguez R, et al. The outflow tract of the heart is recruited from a novel heart-forming field. *Dev Biol.* 2001;238(1):97–109.
13. Waldo KL, Kumiski DH, Wallis KT, et al. Conotruncal myocardium arises from a secondary heart field. *Development.* 2001;128(16):3179–3188.
14. Kelly RG, Brown NA, Buckingham ME. The arterial pole of the mouse heart forms from Fgf10-expressing cells in pharyngeal mesoderm. *Dev Cell.* 2001;1(3):435–440.

15. Harris IS, Black BL. Development of the endocardium. *Pediatr Cardiol.* 2010; 31(3):391–399.
16. Lough J, Sugi Y. Endoderm and heart development. *Dev Dyn.* 2000;217(4):327–342.
17. Kattman SJ, Huber TL, Keller GM. Multipotent flk-1+ cardiovascular progenitor cells give rise to the cardiomyocyte, endothelial, and vascular smooth muscle lineages. *Dev Cell.* 2006;11(5):723–732.
18. Cai CL, Liang X, Shi Y, et al. Isl1 identifies a cardiac progenitor population that proliferates prior to differentiation and contributes a majority of cells to the heart. *Dev Cell.* 2003;5(6):877–889.
19. Verzi MP, McCulley DJ, De Val S, Dodou E, Black BL. The right ventricle, outflow tract, and ventricular septum comprise a restricted expression domain within the secondary/anterior heart field. *Dev Biol.* 2005;287(1):134–145.
20. Ferdous A, Caprioli A, Iacovino M, et al. Nkx2-5 transactivates the Ets-related protein 71 gene and specifies an endothelial/endocardial fate in the developing embryo. *Proc Natl Acad Sci USA.* 2009;106(3):814–819.
21. Milgrom-Hoffman M, Harrelson Z, Ferrara N, Zelzer E, Evans SM, Tzahor E. The heart endocardium is derived from vascular endothelial progenitors. *Development.* 2011;138(21):4777–4787.
22. Manasek FJ. Embryonic development of the heart I: a light and electron microscope study of myocardial development in the early chick embryo. *J Morphol.* 1968; 125:329–366.
23. Manasek FJ, Burnside MB, Waterman RE. Myocardial cell shape change as a mechanism of embryonic heart looping. *Dev Biol.* 1972;29:349–371.
24. Christoffels VM, Habets PE, Franco D, et al. Chamber formation and morphogenesis in the developing mammalian heart. *Dev Biol.* 2000;223:266–278.
25. Lyons I, Parsons LM, Hartley L, et al. Myogenic and morphogenetic defects in the heart tubes of murine embryos lacking the homeo box gene Nkx2-5. *Genes Dev.* 1995; 9:1654–1666.
26. Takeda S, Yonekawa Y, Tanaka Y, Okada Y, Nonaka S, Hirokawa N. Left-right asymmetry and kinesin superfamily protein KIF3A: new insights in determination of laterality and mesoderm induction by kif3A-/- mice analysis. *J Cell Biol.* 1999;145:825–836.
27. Takeuchi JK, Lou X, Alexander JM, et al. Chromatin remodelling complex dosage modulates transcription factor function in heart development. *Nat Commun.* 2011;2:187.
28. He A, Kong SW, Ma Q, Pu WT. Co-occupancy by multiple cardiac transcription factors identifies transcriptional enhancers active in heart. *Proc Natl Acad Sci USA.* 2011;108(14): 5632–5637.
29. Moorman AF, Christoffels VM, Anderson RH, van den Hoff MJ. The heart-forming fields: one or multiple? *Philos Trans R Soc Lond B Biol Sci.* 2007;362(1484):1257–1265.
30. Vincent SD, Buckingham ME. How to make a heart: the origin and regulation of cardiac progenitor cells. *Curr Top Dev Biol.* 2010;90:1–41.
31. Rana MS, Christoffels VM, Moorman AF. A molecular and genetic outline of cardiac morphogenesis. *Acta Physiol (Oxf).* 2013;207(4):588–615.
32. Francou A, Saint-Michel E, Mesbah K, et al. Second heart field cardiac progenitor cells in the early mouse embryo. *Biochim Biophys Acta.* 2013;1833(4):795–798.
33. Cohen ED, Wang Z, Lepore JJ, et al. Wnt/beta-catenin signaling promotes expansion of Isl-1-positive cardiac progenitor cells through regulation of FGF signaling. *J Clin Invest.* 2007;117(7):1794–1804.
34. Kwon C, Arnold J, Hsiao EC, Taketo MM, Conklin BR, Srivastava D. Canonical Wnt signaling is a positive regulator of mammalian cardiac progenitors. *Proc Natl Acad Sci USA.* 2007;104(26):10894–10899.
35. Ai D, Fu X, Wang J, et al. Canonical Wnt signaling functions in second heart field to promote right ventricular growth. *Proc Natl Acad Sci USA.* 2007;104(22):9319–9324.

36. Galli D, Domínguez JN, Zaffran S, Munk A, Brown NA, Buckingham ME. Atrial myocardium derives from the posterior region of the second heart field, which acquires left-right identity as Pitx2c is expressed. *Development.* 2008;135(6):1157–1167.
37. Yutzey KE, Rhee JT, Bader D. Expression of the atrial-specific myosin heavy chain AMHC1 and the establishment of anteroposterior polarity in the developing chicken heart. *Development.* 1994;120:871–883.
38. Xavier-Neto J, Neville CM, Shapiro MD, et al. A retinoic acid-inducible transgenic marker of sino-atrial development in the mouse heart. *Development.* 1999;126:2677–2687.
39. Bertrand N, Roux M, Ryckebüsch L, et al. Hox genes define distinct progenitor sub-domains within the second heart field. *Dev Biol.* 2011;353(2):266–274.
40. Chin AJ, Saint-Jeannet JP, Lo CW. How insights from cardiovascular developmental biology have impacted the care of infants and children with congenital heart disease. *Mech Dev.* 2012;129(5–8):75–97.
41. Stevens KN, Hakonarson H, Kim CE, et al. Common variation in ISL1 confers genetic susceptibility for human congenital heart disease. *PLoS One.* 2010;5(5):e10855.
42. Lindsay EA, Su H, Morishima M, et al. Tbx1 haploinsufficiency in the DiGeorge syndrome region causes aortic arch defects in mice. *Nature.* 2001;410:97–101.
43. Jerome LA, Papaioannou VE. DiGeorge syndrome phenotype in mice mutant for the T-box gene, Tbx1. *Nat Genet.* 2001;27:286–291.
44. Merscher S, Funke B, Epstein JA, et al. TBX1 is responsible for cardiovascular defects in velo-cardio-facial/DiGeorge syndrome. *Cell.* 2001;104:619–629.
45. van den Boogaard M, Wong LY, Tessadori F, et al. Genetic variation in T-box binding element functionally affects SCN5A/SCN10A enhancer. *J Clin Invest.* 2012;122(7): 2519–2530.
46. Christoffels VM, Hoogaars WM, Tessari A, Clout DE, Moorman AF, Campione M. T-box transcription factor Tbx2 represses differentiation and formation of the cardiac chambers. *Dev Dyn.* 2004;229(4):763–770.
47. Hoogaars WM, Tessari A, Moorman AF, et al. The transcriptional repressor Tbx3 delineates the developing central conduction system of the heart. *Cardiovasc Res.* 2004; 62(3):489–499.
48. Singh R, Horsthuis T, Farin HF, et al. Tbx20 interacts with smads to confine tbx2 expression to the atrioventricular canal. *Circ Res.* 2009;105(5):442–452.
49. Schlueter J, Brand T. Epicardial progenitor cells in cardiac development and regeneration. *J Cardiovasc Transl Res.* 2012;5(5):641–653.
50. Gittenberger-de Groot AC, Vrancken Peeters MP, Mentink MM, Gourdie RG, Poelmann RE. Epicardium-derived cells contribute a novel population to the myocardial wall and the atrioventricular cushions. *Circ Res.* 1998;82(10):1043–1052.
51. Vrancken Peeters MP, Gittenberger-de Groot AC, Mentink MM, Poelmann RE. Smooth muscle cells and fibroblasts of the coronary arteries derive from epithelial-mesenchymal transformation of the epicardium. *Anat Embryol (Berl).* 1999;199(4): 367–378.
52. Katz TC, Singh MK, Degenhardt K, et al. Distinct compartments of the proepicardial organ give rise to coronary vascular endothelial cells. *Dev Cell.* 2012;22(3):639–650.
53. Pérez-Pomares JM, Phelps A, Sedmerova M, et al. Experimental studies on the spatiotemporal expression of WT1 and RALDH2 in the embryonic avian heart: a model for the regulation of myocardial and valvuloseptal development by epicardially derived cells (EPDCs). *Dev Biol.* 2002;247(2):307–326.
54. von Gise A, Zhou B, Honor LB, Ma Q, Petryk A, Pu WT. WT1 regulates epicardial epithelial to mesenchymal transition through β-catenin and retinoic acid signaling pathways. *Dev Biol.* 2011;356(2):421–431.
55. Chen T, Chang TC, Kang JO, et al. Epicardial induction of fetal cardiomyocyte proliferation via a retinoic acid-inducible trophic factor. *Dev Biol.* 2002;250(1):198–207.

56. Pennisi DJ, Ballard VL, Mikawa T. Epicardium is required for the full rate of myocyte proliferation and levels of expression of myocyte mitogenic factors FGF2 and its receptor, FGFR-1, but not for transmural myocardial patterning in the embryonic chick heart. *Dev Dyn.* 2003;228(2):161–172.
57. Stuckmann I, Evans S, Lassar AB. Erythropoietin and retinoic acid, secreted from the epicardium, are required for cardiac myocyte proliferation. *Dev Biol.* 2003;255(2): 334–349.
58. Merki E, Zamora M, Raya A, et al. Epicardial retinoid X receptor alpha is required for myocardial growth and coronary artery formation. *Proc Natl Acad Sci USA.* 2005; 102(51):18455–18460.
59. Liu J, Bressan M, Hassel D, et al. A dual role for ErbB2 signaling in cardiac trabeculation. *Development.* 2010;137(22):3867–3875.
60. Peshkovsky C, Totong R, Yelon D. Dependence of cardiac trabeculation on neuregulin signaling and blood flow in zebrafish. *Dev Dyn.* 2011;240(2):446–456.
61. Staudt DW, Liu J, Thorn KS, Stuurman N, Liebling M, Stainier DY. High-resolution imaging of cardiomyocyte behavior reveals two distinct steps in ventricular trabeculation. *Development.* 2014;141(3):585–593.
62. Lee KF, Simon H, Chen H, Bates B, Hung MC, Hauser C. Requirement for neuregulin receptor erbB2 in neural and cardiac development. *Nature.* 1995;378:394–398.
63. Hertig CM, Kubalak SW, Wang Y, Chien KR. Synergistic roles of neuregulin-1 and insulin-like growth factor-I in activation of the phosphatidylinositol 3-kinase pathway and cardiac chamber morphogenesis. *J Biol Chem.* 1999;274:37362–37369.
64. Lai D, Liu X, Forrai A, et al. Neuregulin 1 sustains the gene regulatory network in both trabecular and nontrabecular myocardium. *Circ Res.* 2010;107(6):715–727.
65. Grego-Bessa J, Luna-Zurita L, del Monte G, et al. Notch signaling is essential for ventricular chamber development. *Dev Cell.* 2007;12(3):415–429.
66. Luxán G, Casanova JC, Martínez-Poveda B, et al. Mutations in the NOTCH pathway regulator MIB1 cause left ventricular noncompaction cardiomyopathy. *Nat Med.* 2013;19(2):193–201.
67. Bruneau BG, Logan M, Davis N, et al. Chamber-specific cardiac expression of Tbx5 and heart defects in Holt-Oram syndrome. *Dev Biol.* 1999;211:100–108.
68. Takeuchi JK, Ohgi M, Koshiba-Takeuchi K, et al. Tbx5 specifies the left/right ventricles and ventricular septum position during cardiogenesis. *Development.* 2003;130(24): 5953–5964.
69. Durocher D, Charron F, Warren R, Schwartz RJ, Nemer M. The cardiac transcription factors Nkx2-5 and GATA-4 are mutual cofactors. *EMBO J.* 1997;16:5687–5696.
70. Hiroi Y, Kudoh S, Monzen K, et al. Tbx5 associates with Nkx2-5 and synergistically promotes cardiomyocytes differentiation. *Nat Genet.* 2001;28:276–280.
71. Calvert PA, Rana BS, Kydd AC, Shapiro LM. Patent foramen ovale: anatomy, outcomes, and closure. *Nat Rev Cardiol.* 2011;8(3):148–160.
72. Snarr BS, O'Neal JL, Chintalapudi MR, et al. Isl1 expression at the venous pole identifies a novel role for the second heart field in cardiac development. *Circ Res.* 2007; 101(10):971–974.
73. Anderson RH, Brown NA, Moorman AF. Development and structures of the venous pole of the heart. *Dev Dyn.* 2006;235(1):2–9.
74. Schott JJ, Benson DW, Basson CT, et al. Congenital heart disease caused by mutations in the transcription factor NKX2-5. *Science.* 1998;281:108–111.
75. Benson DW, Silberbach GM, Kavanaugh-McHugh A, et al. Mutations in the cardiac transcription factor NKX2.5 affect diverse cardiac developmental pathways. *J Clin Invest.* 1999;104(11):1567–1573.
76. Goldmuntz E, Geiger E, Benson DW. NKX2.5 mutations in patients with tetralogy of fallot. *Circulation.* 2001;104:2565–2568.

77. Garg V, Kathiriya IS, Barnes R, et al. GATA4 mutations cause human congenital heart defects and reveal an interaction with TBX5. *Nature*. 2003;424:443–447.
78. de Vlaming A, Sauls K, Hajdu Z, et al. Atrioventricular valve development: new perspectives on an old theme. *Differentiation*. 2012;84(1):103–116.
79. Camenisch TD, Biesterfeldt J, Brehm-Gibson T, Bradley J, McDonald JA. Regulation of cardiac cushion development by hyaluronan. *Exp Clin Cardiol*. 2001;6(1):4–10.
80. Mjaatvedt CH, Yamamura H, Capehart AA, Turner D, Markwald RR. The Cspg2 gene, disrupted in the hdf mutant, is required for right cardiac chamber and endocardial cushion formation. *Dev Biol*. 1998;202(1):56–66.
81. Goddeeris MM, Rho S, Petiet A, et al. Intracardiac septation requires hedgehog-dependent cellular contributions from outside the heart. *Development*. 2008;135(10):1887–1895.
82. Briggs LE, Phelps AL, Brown E, et al. Expression of the BMP receptor Alk3 in the second heart field is essential for development of the dorsal mesenchymal protrusion and atrioventricular septation. *Circ Res*. 2013;112(11):1420–1432.
83. Bax NA, Bleyl SB, Gallini R, et al. Cardiac malformations in Pdgfralpha mutant embryos are associated with increased expression of WT1 and Nkx2.5 in the second heart field. *Dev Dyn*. 2010;239(8):2307–2317.
84. Keyte A, Hutson MR. The neural crest in cardiac congenital anomalies. *Differentiation*. 2012;84(1):25–40.
85. Kirby ML, Gale TF, Stewart DE. Neural crest cells contribute to normal aorticopulmonary septation. *Science*. 1983;220(4601):1059–1061.
86. Anderson RH, Chaudhry B, Mohun TJ, et al. Normal and abnormal development of the intrapericardial arterial trunks in humans and mice. *Cardiovasc Res*. 2012;95(1):108–115.
87. Poelmann RE, Mikawa T, Gittenberger-de Groot AC. Neural crest cells in outflow tract septation of the embryonic chicken heart: differentiation and apoptosis. *Dev Dyn*. 1998;212(3):373–384.
88. van den Hoff MJ, Moorman AF, Ruijter JM, et al. Myocardialization of the cardiac outflow tract. *Dev Biol*. 1999;212(2):477–490.
89. Jiang X, Rowitch DH, Soriano P, McMahon AP, Sucov HM. Fate of the mammalian cardiac neural crest. *Development*. 2000;127(8):1607–1616.
90. Boukens BJ, Christoffels VM. Electrophysiological patterning of the heart. *Pediatr Cardiol*. 2012;33(6):900–906.
91. Christoffels VM, Smits GJ, Kispert A, Moorman AF. Development of the pacemaker tissues of the heart. *Circ Res*. 2010;106(2):240–254.
92. Arnolds DE, Liu F, Fahrenbach JP, et al. TBX5 drives Scn5a expression to regulate cardiac conduction system function. *J Clin Invest*. 2012;122(7):2509–2518.
93. Tartaglia M, Mehler EL, Goldberg R, et al. Mutations in PTPN11, encoding the protein tyrosine phosphatase SHP-2, cause Noonan syndrome. *Nat Genet*. 2001;29:465–468.
94. Yagi H, Furutani Y, Hamada H, et al. Role of TBX1 in human del22q11.2 syndrome. *Lancet*. 2003;362:1366–1373.
95. Basson CT, Bachinsky DR, Lin RC, et al. Mutations in human cause limb and cardiac malformation in Holt-Oram syndrome. *Nat Genet*. 1997;15:30–34.
96. Li QY, Newbury-Ecob RA, Terrett JA, et al. Holt-Oram syndrome is caused by mutations in TBX5, a member of the Brachyury (T) gene family. *Nat Genet*. 1997;15:21–35.

CHAPTER THREE

# MicroRNA in Pulmonary Vascular Disease

**Alex M.K. Rothman, Timothy J.A. Chico, Allan Lawrie**
Department of Cardiovascular Science, University of Sheffield, Sheffield, United Kingdom

## Contents



## Abstract

MicroRNA (miRNA) are short noncoding RNA that regulate gene expression by inhibiting translation or promoting degradation of target mRNA. miRNA are key regulators of a wide range of cellular processes and their discovery has revolutionized our understanding of gene regulatory networks. Pulmonary arterial hypertension (PAH) is a debilitating and fatal disease characterized by remodeling of pulmonary arteries and right heart failure. Factors including sustained pulmonary vasoconstriction, inflammation, and altered cellular signaling pathways drive disease through pulmonary artery endothelial dysfunction, smooth muscle cell proliferation, and the recruitment of circulating cells. miRNA have been shown to regulate many of the key drivers of pathology,

*Progress in Molecular Biology and Translational Science*, Volume 124
ISSN 1877-1173
http://dx.doi.org/10.1016/B978-0-12-386930-2.00003-3

yet the role of only a limited number of miRNA has been recognized in PAH. Investigation of the diverse regulatory functions of miRNA offers the potential to further understanding of the cellular pathology of PAH and to provide much needed diagnostic and therapeutic strategies. This review focuses on recent advances in the investigation of miRNA in PAH.

# 1. INTRODUCTION

Recent discoveries in regulation of gene expression have added complexity to our understanding of cellular biology. Human disease was once thought to arise from dysregulation of protein-coding genes and the protein–protein interaction of their products. However, epigenetic modifications of the genome and posttranscriptional regulation of gene expression by noncoding RNA are now recognized as powerful regulatory mechanisms that significantly influence health and disease. Since the first description of their regulatory capacity in the early 1990s, short noncoding RNA termed microRNA (miRNA) have emerged as key regulators of a range of adaptive cellular functions. The development of high-throughput techniques to quantify their expression has led to significant advances in the understanding of the role of miRNA in cellular pathology.

## 1.1. Pulmonary arterial hypertension

Pulmonary arterial hypertension (PAH) is a devastating and debilitating disease. Current clinical therapies provide only a modest reduction in mortality and patients with the disease experiencing life-limiting breathlessness and heart failure. Meta-analysis of randomized placebo controlled trail data suggest that PAH-specific therapies provide an average absolute risk reduction of 2.3% over 14.3 weeks.[1,2] Despite these treatments outcomes remain poor and mortality within groups treated with PAH-specific therapy was 1.54% over the same period.[2] The pathophysiological mechanisms underlying PAH are complex and incompletely understood.[3] The cellular processes driving the characteristic vascular remodeling seen in PAH has been likened to neoplasia with prominent expansion of pulmonary arterial smooth muscle cells (PASMCs) and pulmonary artery endothelial cells (PAEC).[4] Factors including inflammation,[5] drugs,[6] and gene mutation of members of the transforming growth factor beta (TGF-β)[7,8] have been implicated in early-disease pathology. Damage to PAEC precedes PASMC proliferation[9,10] leading to a proproliferative, antiapoptotic state, and to the subsequent loss of the endothelial monolayer.[11,12] The resultant endothelial dysfunction alters the balance of cytokines,[5,13] growth factors,[1,14] and vasoactive substances[3,15] within the pulmonary vasculature causing the vasoconstriction, and PAEC and PASMC proliferation.[4,16] Proliferation of the cellular components of the vessel wall in small- and medium-sized pulmonary arteries results in the formation of the concentric and plexiform lesions that are characteristic of PAH.

## 1.2. Animals models of PAH

Animal models, although imperfect in their replication of human disease, are essential tools for the investigation of cellular pathology, identification of therapeutic targets, and the testing of new clinical therapies in PAH. To increase confidence in translation, investigators use a range of models to examine different characteristics of disease pathology. The models discussed in this chapter are briefly outlined below.

*Chronic hypoxia*: Hypoxic exposure (usually to 10% oxygen) of the rat or mouse (weaker model) leads to vasoconstriction, which can stimulate muscularization of the small pulmonary arteries, thickening, and fibrosis of the proximal pulmonary arteries and a rise in mean pulmonary artery pressure over a 2-week period. These changes are associated with right ventricular hypertrophy but not with right heart failure.[17]

*Monocrotaline*: The toxic pyrrolizidine alkaloid monocrotaline is metabolized by hepatic oxidases to monocrotaline pyrrole. The active compound causes vascular injury through endothelial damage and inflammation resulting in pulmonary hypertension and subsequent right heart failure in rats.[17]

*Sugen5416 combined with Hypoxia (SuHx)*: Vascular endothelial growth factor is central to the maintenance of vascular endothelial integrity.[18] Administration of the VEGF inhibitor Sugen5416 in the presence of chronic hypoxia results in severe pulmonary hypertension associated with pulmonary vascular changes somewhat similar to human disease and right heart failure.[18]

## 1.3. MicroRNA

RNA is a unique informational molecule. Linear base sequences encode genetic information, while intricate secondary and tertiary structures allow RNA to influence a range of cellular processes including transcription, translation, splicing, cellular localization, and mRNA degradation. Since the early 1990s, new light has been shed on the regulatory functions of noncoding RNA. The discovery that the short noncoding RNA product of the *LIN-4* gene showed sequence complementarity to regions in the 3′UTR of LIN-14 and that expression of LIN-4 and LIN-14 was inversely related lead to the hypothesis (now confirmed) that noncoding RNAs regulate gene expression through Watson–Crick base-pairing interactions.[19,20] Over 1800 human miRNA have been identified[21] and sequence complementarity-based target predictions suggest as many as 60% of human proteins may be regulated by miRNA binding.[22,23] miRNA regulate expression of multiple mRNA transcripts providing a mechanism by which small changes in expression of a single miRNA influence multiple cellular signaling pathways simultaneously.

The first association between miRNA and human disease was the observation that miR-15 and miR-16 expression was reduced in patients with B-cell leukemia.[24] Since then, significant advances have been made in understanding the role of miRNA in health and disease. miRNA regulate cardiac development, modulate established cardiovascular risk factors such as hypercholesterolemia,[25,26] and are implicated in cardiovascular diseases such as atherosclerosis,[27,28] heart failure,[29,30] and pulmonary hypertension.[31–33] The area of miRNA-based therapeutics has also shown great potential. Exciting preclinical studies have shown improvement of disease phenotype employing different strategies to increase[32] or decrease levels of miRNA.[30–32] These promising results have led to the first human trial of an miRNA-targeting therapy in which patients with Hepatitis C were treated with a locked nucleic acid antagomir to miR-122, successfully reducing levels of the virus.[34]

PAH is a biologically complex disease driven by factors including hypoxia, inflammation, and altered cellular signaling. miRNA are key regulators of many of these pathways, yet the role of only a small number has been investigated in PAH. The diverse cellular processes underlying PAH and the pleotropic, cooperative effects of miRNA suggest that continued investigation will provide further insights into disease with potential translation applications as biomarkers of disease and even as potential therapeutic agents.

In this chapter, we describe the biogenesis, regulation, and cellular biology of miRNA and review current understanding of their role in the pathology of PAH.

## 2. miRNA BIOGENESIS AND FUNCTION

Mature miRNA are processed from precursor molecules (pre-miRNA), which are transcribed in an RNA polymerase II-dependent manner by two distinct mechanisms (Fig. 3.1). Transcription initiated at a miRNA-specific promoter forms a pri-miRNA (primary-miRNA) which is processed to a ~70 nucleotide pre-miRNA (precursor-miRNA) by Drosha, a member of the RNase III family. Alternately, miRNA may be produced from a mirtron, a miRNA-encoding sequence located between exons of a host protein-coding sequence. Transcription initiated at a non-miRNA-specific promoter transcribes the protein-coding exons and intervening mirtron in a single pre-mRNA sequence. The pre-mRNA is spliced, debranched, and folded to form a pre-miRNA. The pre-miRNA transcript, formed via either mechanism, is bound by Exportin 5 and translocated to the cytoplasm where cleavage by the RNase III Dicer produces a ~21-nucleotide miRNA–miRNA* duplex. A single strand of this duplex, the mature miRNA, is preferentially incorporated into the mRNA silencing

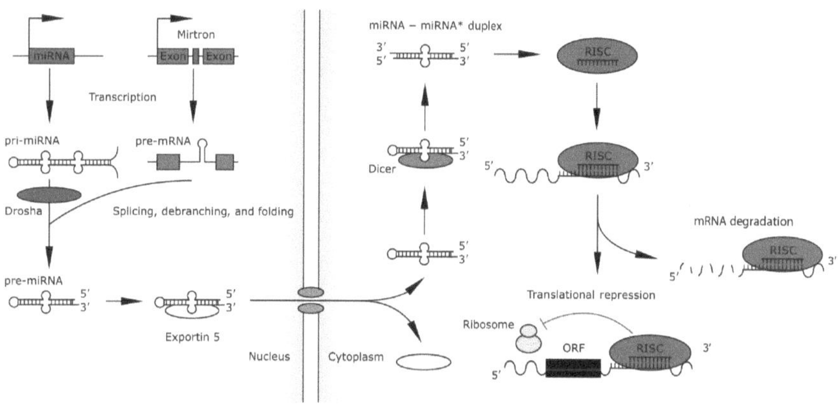


**Figure 3.1** Biogenesis and effector machinery of miRNA. MicroRNA are transcribed from miRNA genes as primary-miRNA (pri-miRNA) or from mirtrons located between exons of protein-coding genes as part of a pre-mRNA. The precursor-miRNA (pre-miRNA) is formed by processing of the pri-miRNA by the Drosha complex or by cleavage of the pre-mRNA by the lariat-debranching enzyme. In association with Exportin 5, the pre-miRNA is exported from the nucleus to the cytoplasm where further processing by Dicer generates a mature miRNA–miRNA* duplex. The single-stranded miRNA is loaded into the RNA-induced silencing complex (RISC). Binding of the RISC-incorporated miRNA to a target mRNA silences gene expression by translational repression or degradation. (See the color plate.)

complex (RISC) and the non-RISC-incorporated strand, the passenger or miRNA* strand, is discarded. Complementary base pairing between the RISC-incorporated miRNA and target mRNA sequences results in down-regulation of the target gene transcript by degradation or transcriptional repression.[35]

## 2.1. Regulation of miRNA expression

Promoter regions of miRNA genes and protein-coding regions share many common features. Therefore, the expression of miRNA varies with time and cell type, dependent on transcription factor binding and epigenetic regulation of the miRNA promoter.[36] Approximately, 50% of miRNA genes are located in noncoding regions of the genome with their transcription regulated by activation and repression of their promoter. Transcriptional regulation of the ~40% of miRNA encoded by mirtrons is more complex. These miRNA are often coexpressed with the mRNA transcript in which they reside; however, the presence of miRNA gene within a host gene does not imply coregulation. As many as 33% of intronic miRNA are expressed from promoters distinct to those regulating the gene in which they are located.[37] miRNA regulated in this manner are located farther from the host gene transcription start site than those using the host gene promoter, a factor that may have evolved to increase speed of pri-miRNA transcription.[37]

Transcription of a single miRNA gene gives rise to a single mature miRNA; however, miRNA-encoding genes are also arranged in groups or clusters. Transcription of a miRNA cluster forms a single pri-miRNA containing several hairpin loops, each of which may be processed to form a mature miRNA.[38] The expression of individual miRNA from this cluster may be altered by transcription initiated at independent promoters[39] or by differential processing of the pri-miRNA transcript.[40]

Activation and repression of promoters is a primary factor dictating miRNA transcription, however, other regulatory mechanisms also affect cellular levels and function. Figure 3.2 illustrates the effects of miRNA networks, processing, stability, and RNA-binding proteins on miRNA function.

## 2.2. MicroRNA/mRNA interaction and target prediction

Interactions between miRNA and target mRNA are primarily mediated through the formation of partial duplexes between nucleotides 2 and 7 of the 5′ end of the miRNA (the seed region) and complimentary sequences within the 3′UTR of the target mRNA.[23] Secondary and tertiary structures

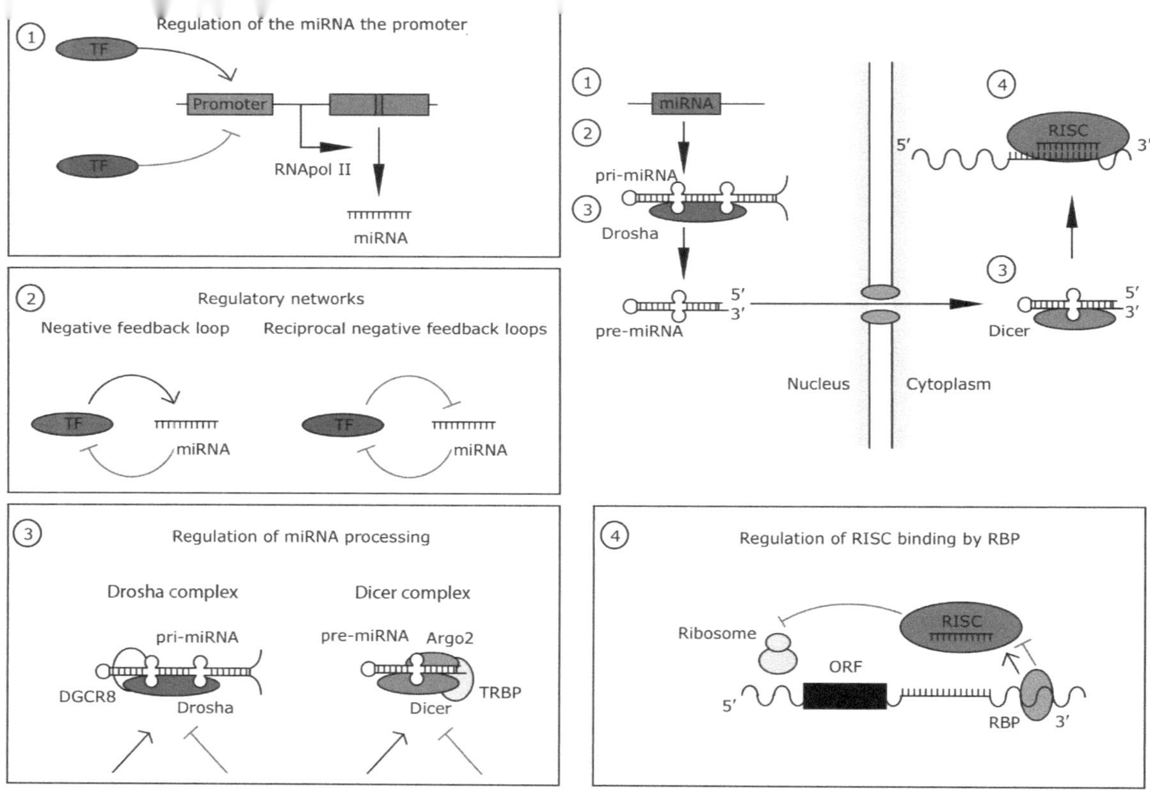


**Figure 3.2** Mechanisms of regulation of miRNA expression and function. (1) Regulation of miRNA transcription is similar to that of protein-coding sequences. Transcription factors bind the miRNA promoter activating (green) or repressing (red) miRNA transcription. The abundance of transcription factors is both tissue and stimulus specific. (2) Feedback loops are central to the regulation of key cellular processes. miRNA are able to directly target mRNA sequences encoding transcription factors regulating their expression. miRNA inhibition of a positive transcription factor (green) forms a negative feedback loop. miRNA inhibition of a negative transcription factor (red) creates a reciprocal negative feedback loop. More complex regulatory networks exist, but are not depicted. (3) Cleavage by Drosha and Dicer are required for production of a mature miRNA. Drosha and Dicer operate in complexes with RNA-binding proteins (RBP). The levels and activity of these proteins and of accessory proteins influences miRNA processing. (4) RBP binding to target mRNA facilitates or inhibits binding of the RISC-incorporated miRNA, thereby altering transcriptional inhibition or degradation. (See the color plate.)

formed by linear RNA sequences lead to flexible imperfect base paring and factors beyond sequence analogy determine functional target interactions *in vivo*. Factors such as conserved AU-rich regions flanking the miRNA target-binding site, proximity of the binding site to other sites for coexpressed miRNA (which leads to cooperative action), proximity of the site to residues pairing to miRNA nucleotides 13–16 (increasing binding stability), positioning within the 3′UTR at least 15 nucleotides from the stop codon, and positioning toward the end of long 3′UTRs[41,42] are employed in addition to simple sequence analogy in algorithms used to predict the targets of individual miRNA. The specificity and sensitivity of target prediction remains imperfect and *in vitro* and *in vivo* testing is required to characterize miRNA–mRNA interactions and functional effects.[43,44]

## 2.3. Cellular localization of miRNA

Cellular compartmentalization of miRNA has significant effects on its interactions and function as for an interaction to take place with a target both must physically interact. The mechanisms determining the cellular localization of miRNA are yet to be fully elucidated.[45] miRNA have been identified within the nucleus and cytoplasm of cells and in the circulation bound within microvesicles, microparticles, and apoptotic bodies or bound to proteins and lipids.[46]

Microvesicles shed from endothelial cells in response to shear stress have been shown to contain miR-143 and miR-145 and to modulate expression of miR-143 and miR-145 target genes in vascular smooth muscle cells in coculture.[27] This highlights both the importance of cellular compartmentalization of miRNA and their potential role as intermediate messengers in cell–cell communication.

# 3. miRNA IN PAH

PAH is a biologically complex disease with a variety of predisposing and underlying factors contributing to onset and progression. Expression analysis in a range of model systems and human tissues has been used to identify miRNA with roles in PAH. Table 3.1 summarizes miRNA implicated in PAH listing potential disease relevant targets.

**Table 3.1** Summary of miRNA linked to PAH pathogenesis in animal models and/or tissue samples from patients with PAH

| miRNA | Model/patient tissue | Change | Stimulus | PAH targets | References |
|---|---|---|---|---|---|
| miR-22 | MCT, Hx | ↓ | | KCN56, NFAT5, PPP2R5E, TNRC6A, ADARB, TGFBR1, TGFBRAP | 47 |
| miR-30c | MCT, Hx | ↓ | | KCN56, NFAT5, TACC1, YWHAZ | 47 |
| Let-7f | MCT, Hx | ↓ | | | 47 |
| miR-322 | MCT, Hx | ↑ | | | 47 |
| miR-451 | MCT, Hx | ↑ | | | 47 |
| miR-21 | MCT, human lung and serum | ↓ | | TACC1, PPP2R5E, RBPSUH | 47 |
| Let-7a | MCT | ↓ | | | 47 |
| miR-143 | Formalin-fixed human lung | ↓(PL) ↑(CL) | BMP-4 | KLF4, smMHC | 48 |
| miR-145 | Formalin-fixed human lung | ↓(PL) ↑(CL) | BMP-4 | KLF4, smMHC | 48 |
| miR-21 | Formalin-fixed human lung | ↑ | | | 48 |
| miR-204 | Formalin-fixed human lung | ↓(PL) ↑(CL) | | | 48 |
| miR-126 | Formalin-fixed human lung | ↓ | | VEGFA | 48 |
| miR-19a/b | Formalin-fixed human lung | ↓ | | | 48 |
| miR-92a | Formalin-fixed human lung | ↑ | | | 48 |
| miR-204 | Human PASMC—explanted | ↓ | STAT3 | SHP2 | 32 |

*Continued*

**Table 3.1** Summary of miRNA linked to PAH pathogenesis in animal models and/or tissue samples from patients with PAH—cont'd

| miRNA | Model/patient tissue | Change | Stimulus | PAH targets | References |
|---|---|---|---|---|---|
| miR-150 | Human plasma | ↓ | KLF2 | | 49 |
| miR-424, miR-503 | Human PAEC | ↓ | Apelin | FGF2, FGFR1 | 33 |
| miR-210 | Human PASMC | ↑ | HIF1α | E2F3 | 50 |
| miR-17 | Promoter analysis | | IL-6, STAT3 | BMPR2 | 51 |
| miR-20a | Promoter analysis | | IL-6, STAT3 | BMPR2 | 51 |
| | SuHx, MCT mouse | ↑ | BMPR2 | RhoB, BMPR2 | 52 |
| miR-17 | | | | P21 | 53 |
| miR-145 | Hypoxic mouse PASMC | ↑ | | SMAD4, SMAD5, KLF-4 | 31 |

PL, plexiform lesion; CL, concentric lesion.

### 3.1. miRNA in animal models of PAH

The hypoxic rat and the monocrotaline-treated rat are two common models of PAH thought to replicate different aspects of its pathology. Using these models, Caruso and colleagues performed a longitudinal study of miRNA expression during PAH development.[47] Over the duration of the study, reduction of the miRNA-processing enzyme Dicer was noted in both models (hypoxic rat from day 2 onward and at day 21 in the monocrotaline rat). Two-channel microarray was used to quantify miRNA expression in whole-lung extracts and identified downregulation of let-7f, miR-22, and miR-30c and upregulation of miR-322 and miR-451 in both models. However, not all identified changes were consistent between models and levels of let-7a and miR-21 were downregulated only in the monocrotaline model. The number of miRNA to be further investigated was rationalized using cell culture. Hypoxic rat PASMCs and PAF were cultured and changes in miRNA levels quantified by qPCR. Let-7f, miR-21, miR-22, and miR-30c showed changes consistent with the *in vivo* findings. The effects of both TGF-β1 and BMP4 on miRNA levels in PASMC were determined by qPCR and showed an increase in miR-421, whereas TGF-β1 alone reduced expression of miR-21, miR-22, miR-30c, and let-7f. Employing bioinformatic target-prediction algorithms, potential targets were identified by examining the 3′UTR of mRNA sequences for matches to 6-mer seed regions of the differentially expressed miRNAs. PPP2R5E, YWHAZ, and TNRC6A, TGFR-B1, KCNJ6, and TACC1 were identified as potential targets, changes which were validated by qPCR in the appropriate animal models. Finally, the levels of the most dysregulated miRs (miR-451 and miR-21) were examined in paraffin-embedded human lung samples and serum samples. Levels of miR-21 were reduced in lung samples and serum from patients with PAH, but levels of miR-451 were unaltered.

Differences highlighted between miRNA levels during the development of PH in the hypoxic and monocrotaline rat perhaps reflect the different disease stimulus in these models. Another potential explanation for the differences between models may be the varied expression of miRNA in different cell types. Whole-lung samples contain a mixture of cells including endothelial cells, smooth muscle, and fibroblasts. The varied contribution of miRNA expression in the varied cell types seen in pulmonary vascular lesions was highlighted in a study by Bockmeyer and colleagues.[48] Plexiform and concentric lesions from formalin sections of patients with PAH and controls were stained *in situ* to quantify expression of miR-143/145 (smooth muscle cell specific) and miR-126 (endothelial cell specific). Levels of miR-143/145 and miR-204 levels were higher in concentric lesions and miR-21 and miR-126 higher in plexiform lesions. These findings suggest that plexiform lesions form from a prominent angiogenic phenotype of cells.[48] The varied and relative contribution of each cell type may explain the disparity in the regulation of miR-21 identified in studies in both rodents and humans.[54,55]

The cellular consequences of miR-21 and miR-204 dysregulation have been further investigated in studies described below.[32,52]

Animal models of PAH provide an excellent platform for the investigation of disease, however, no one model faithfully replicates human disease.[17] The analysis of miRNA expression cells derived from patients with PAH potentially offers more direct relevance to, and a more reliable means of identifying biomarkers, human disease.

### 3.2. Ex vivo human pulmonary artery smooth muscle cells: miR-204 and STAT target signaling

Courboulin and colleagues used qPCR arrays to examine miRNA expression in PASMC from explanted lungs of patients with PAH.[32] Dysregulation of seven miRNAs was identified and subsequently validated in lung biopsies from patients with PAH. Downregulation of miR-204 both correlated with severity of disease on biopsy and performed as a circulating biomarker of PAH when measured in Buffy coat cells. The downregulation of miR-204 was shown to reduce the direct inhibition of SHP2 by miR-204 thus activating Src kinase and nuclear factor of activated T cells (NFAT) to reduce apoptosis and increase proliferation of PASMC in culture. The reduction of miR-204 seen in humans was also apparent in the monocrotaline rat model of PAH with miR-204 levels gradually reducing as PAH progressed. Finally, administration of synthetic miR-204 by intratracheal nebulization normalized levels of miR-204, SHP2, and STAT3 and reduced mPAP and right ventricular hypertrophy. This study was the first to suggest that a single miRNA may be a viable therapeutic target for treatment of PAH.

### 3.3. miRNA in human plasma: miR-150 as a biomarker of PAH

miRNA are present in a range of blood fractions and have been tested as diagnostic biomarkers in a variety of diseases.[28,32,49,56] In search of clinical biomarkers of disease severity, Rhodes and colleagues examined expression of miRNA in plasma from eight patients newly diagnosed with PAH and eight healthy volunteers.[49] Total RNA extracted from plasma was pooled for microarray analysis of miRNA levels. 58 miRNAs were differentially expressed between groups. In contrast to previous studies, miR-204, miR-17, and miR-92a were not differentially expressed, perhaps highlighting differences in miRNA expression between different tissues and different blood fractions. miR-150 showed the greatest downregulation in PAH, validated by qPCR in two independent cohorts. Reduced expression of miR-150 predicted 2-year survival and correlated with disease severity. miR-150 was identified in circulating microvesicles and lymphocytes; however, levels of miR-150 in lymphocytes were thought to account for only 6% of variability of miR-150 levels and circulating lymphocyte number did not predict disease survival. Finally, reduced levels of miR-150 and KLF2, a recently identified regulator of miR-150, were identified in the monocrotaline rat model of PAH. The source of the miR-150-depleted microvesicles in PAH remains to be determined; however, it was suggested that miR-150 may be shed from the lungs or from circulating peripheral blood mononuclear cells. Further studies are required to determine the cellular origin of the dysregulated miR-150 to understand its temporal expression and to elucidate is role in the pathology of PAH.

## 3.4. Pathway-based investigation of miRNA in PAH

A number of signaling pathways are thought to underpin the cellular proliferation characteristic of PAH. Identifying miRNA related to these pathways is another strategy to identify miRNA that contribute to or regulate the disease. This approach provides insight into the specific role of miRNA related directly to the signaling pathways investigated.

## 3.5. miR-424/503 mediate secondary effects of the Apelin–Apelin receptor axis

Recent studies have described an emerging role for the Apelin–Apelin receptor signaling axis in PAH[57] and Apelin expression is reduced in serum and pulmonary microvascular endothelium of patients with PAH.[58] Kim and colleagues identified a reduction in Apelin expression in PAEC cells from patients with PAH.[33]

Augmentation of Apelin in healthy human PAEC leads to a proproliferative response; however, in PAEC from patients with PAH, increased Apelin levels inhibit proliferation. This finding has led to the hypothesis that reduced Apelin in PAH triggers a secondary signaling cascade that increases proliferation of human PAEC.

A range of angiogenic growth factors has been investigated as potential downstream signaling targets of Apelin, including FGF2. Apelin and FGF2 mRNA levels are inversely related in both control and patient PAECs. In cell culture, knockdown of Apelin results in increased FGF2 levels, while Apelin overexpression reduces FGF2. Furthermore, levels of FGF2 are increased in Apelin null mice compared to controls.[33]

In control human PAECs, knockdown of the miRNA-processing enzyme ARGO2 increases FGF2.[33] When performed in PAECs from patients with PAH, knockdown of ARGO2 had no effect. These experiments indicate that FGF2 regulation is mediated by miRNA in human PAECs and that this interaction may be compromised in PAH. miRNA microarray studies and subsequent qPCR of samples from human PAECs with reduced expression of Apelin or the Apelin receptor identified 14 downregulated miRNA. *In silico* analysis matched two of these miRNA to the 3′UTR of FGF2: miR-424 and miR-503. A luciferase reporter construct of the miR-424 and miR-503 promoters showed response to Apelin overexpression, and a luciferase reporter construct of the 3′UTRs of FGF2 and FGFR1 were shown to respond to miR-424 and miR-503 levels. Mutation of the promoter sequence of each construct abrogated the observed effect, showing FGF2 and FGFR1 are direct targets of miR-424 and miR-503. ERK1/2 are downstream effectors of the FGF2–FGFR1 axis. Anti-miR transfection of both miR-424 and miR-503 leads to a significant increase in ERK1/2 phosphorylation.[33]

The functional role of miR-424 and miR-503 in human PAECs was examined in cell culture. Overexpression resulted in reduced proliferation and cell-cycle arrest at the G0/G1 stage, an effect reversed by FGFR1 transfection and FGF2 stimulation.[33]

Transfection of both miR-424 and miR-503 into human PASMC did not alter proliferation. The presence of human PAECs led to increased proliferation of human PASMCs in coculture. Proliferation was reduced in miR-424- and miR-503-transfected human PAECs and abrogated by FGF2 overexpression. Conditioned medium from Apelin knockdown human PAECs induced proliferation of human PASMCs, an effect blocked by overexpression of miR-424 and miR-503.

The role of miR-424 and miR-503 was finally examined in the monocrotaline and Sugen/hypoxia rat models of PAH. Expression of Apelin, miR-424, and miR-503 were decreased and FGF2 and FGFR1 increased in lung and lung endothelial cells compared to control animals. The therapeutic potential of replacing miR-424 and miR-503 in these models was investigated in both the monocrotaline (prevention and rescue) and Sugen hypoxia (rescue) model. Intranasal delivery of a lentiviral vector expressing miR-424 and miR-503 resulted in reduced right ventricular systolic pressure (RVSP) and right ventricular hypertrophy; a reduction in cellular proliferation and arterial obliteration; and reduced expression of FGF2 and FGFR1 in all models.[33]

These data highlight the critical role of miRNA in signaling pathways related to PAH and the potential of focused investigation of miRNA related to specific signaling pathways to provide novel insight into disease.

### 3.6. Hypoxia miR-210 and HIF1α

Hypoxia is both a driving factor in, and a consequence of, PAH. To identify hypoxia-regulated miRNA, Gou and colleagues exposed human PASMC to hypoxic conditions.[50] Hypoxia upregulated miR-210, a finding replicated in lung tissue from hypoxic mice. Human PASMC exposed to hypoxic conditions were used to identify initiators of miR-210 expression and downstream targets in human PASMC. HIF1α was shown to upregulate miR-210 expression and led to apoptosis of human PASMC via the transcription factor E2F3, a direct target of miR-210.

### 3.7. miRNA, inflammation, and BMPR2 signaling

The miR-17–92 cluster was implicated in a range of cardiovascular diseases[5,28,53,59] prior to investigation of its role in PAH. The miR-17–92 cluster are coexpressed in response to transcription factor binding at a single promoter, resulting in the transcription of a single pre-miRNA that is further processed and cleaved to generate the mature miRNAs. Brock and colleagues identified the miR-17–92 cluster as potential modulators of BMPR2 signaling by analyzing the sequence of the BMPR2 promoter and matching this to known miRNA sequences.[51] Using a BMPR2 reporter in HEK293 cells, it was shown that increased miR-17 and miR-20a markedly reduced BMPR2 levels and that the miR-17–92 cluster is induced by IL-6 signaling via STAT3.[51]

Several chemistries have been employed to stabilize and increase cellular permeability of short RNA sequences to aid cellular uptake and prolong half-life. The impact of a cholesterol-modified antagomir to miR-20a was investigated as a potential therapy for PAH. Anti-miR-20a transfection restored functional BMPR2 signaling in human PASMCs and intraperitoneal administration of anti-miR-20a increased BMPR2 levels and reduced vascular remodeling in lung tissue of hypoxic mice, but had no effect on hemodynamic parameters of PH.[60]

miR-21 expression is increased in the distal small pulmonary arteries of the hypoxic mouse.[54] Downregulation of the potential miR-21 targets BMPR2, WWP1, SATB1, and YOD1 were observed in the same tissue. Overexpression and sequestration of miR-21 increased and decreased proliferation of human PASMC, respectively, with modulation of the miR-21 targets PCNA, Cyclin D1, and Bcl-xL. Intratracheal administration of unaltered anti-miR-21 resulted in reduction in RVSP and vascular remodeling in the hypoxic mouse model of PH.

Parikh and colleagues undertook a network biology study to identify miRNA implicated in PAH.[52] Using a list of PH implicated genes and sequence-based target-prediction algorithms, they identified miR-21 as a potential PAH-related miRNA with targets in the BMPR2 and Rho/Rho-kinase-signaling pathways as well as targets related to hypoxia and inflammation. miR-21 was upregulated by both hypoxia and BMPR2 signaling in human PASMC. Targets validated by qPCR included RhoB and Rho-kinase targets, which would be consistent with decreased angiogenesis, and BMPR2; a finding which indicates that miR-21 may form a reciprocal feedback loop with BMPR2 signaling. miR-21 itself was shown to be upregulated in rodent models of PH and lung samples from patients with PAH. Finally, miR-21 null mice were found to have increased RhoB and Rho-kinase expression and an exaggerated PH response to hypoxia.

Pullamsetti and colleagues investigated the effect of antagomirs to the miR-17–92 cluster and miR-21 on PAH and right heart function in the hypoxic mouse and monocrotaline rat models of PAH. Anti-miR-17, -21, and -92a all reduced muscularization of pulmonary arteries in both models and anti-miR-17 and anti-miR-21 reduced RVSP; however, only anti-miR-17 reduced parameters of right heart dysfunction. Levels of miR-17 targets and p21 were altered by administration of anti-miR-17, in both the animal models and human PASMC.[53]

miR-143/145 is considered to be vascular smooth muscle cell specific.[48] Expression of miR-143/145 is driven by TGF-β signalling[61,62] and modulates cell phenotype through KLF4 and myocardin.[63] The role of miR-143/145 was examined by Caruso and colleagues who identified upregulation of miR-145 in the lung and right ventricle of hypoxic mice.[31] To test the role of miR-145 in PAH miR-145$^{-/-}$ mice were exposed to hypoxia, resulting in no change in miR-145 expression and protection from development of PAH. Potential targets of miR-145 SMAD4, SMAD 5, and KLF4 were identified as above and confirmed by qPCR and Western blot. To identify miR-145 targets in PAH, mRNA from wild-type and miR-$145^{-/-}$ mice was used in global transcriptional microarray analysis and the results used for DAVID pathway analysis. qPCR-validated targets were categorized into five groups: inhibitors of Wnt signaling (WIF1, FRZB, DAB2), regulation of actin cytoskeleton (FSCN, TTN), transcriptional regulation (CAMK2A), cell adhesion (ITGBL1, CTGF), and endothelial function (ACE, ANGPTL4). The therapeutic utility of reducing miR-145 levels in PAH was tested in the hypoxic mouse. A locked nucleic acid anti-miR-145 was given subcutaneously to 8-week-old C57Bl6 mice at day 0 and day 8 of hypoxia. This resulted in reduction of RVSP and vessel remodeling, but did not alter right ventricular hypertrophy. Finally, the effect of BMPR2 signaling on mir-145 expression was examined in hPASMC. Knockdown of BMPR2 led to overexpression of miR-145; however, in the reciprocal experiment, an increase in miR-145 failed to have any effect on BMPR2 signaling suggesting that the increase of miR-145 is a downstream effect of reduced BMPR2 signaling.

These data demonstrate the utility of pathway-based exploration in the identification of PAH-related miRNA and the first evidence of the efficacy of anti-miRNA therapies for treatment of PAH.

## 4. CONCLUSIONS AND FUTURE DIRECTIONS

The discovery and functional characterization of miRNA has added greatly to our understanding of the cellular biology and their ubiquitous expression and impact on diverse signaling pathways supports a role as central regulators of cell function. Dysregulation of miRNA expression has been identified up- and downstream of key events in the pathogenesis of PAH and manipulation of specific miRNAs has shown potential therapeutic benefit in model systems. Network analysis and bioinformatics techniques promise to facilitate advancement of the field and the first human trials of miRNA-based therapies in areas away from PAH show exciting potential. However, the cooperative and interactive nature of miRNA function mean that predicting the full implications of manipulating levels *in vivo* is difficult and concerns of unexpected or "off-target" effects provide a significant barrier that must overcome before the full therapeutic potential may be realized.

The early translational impact of miRNA research in PAH may be in areas distant to miRNA-based therapies. Biomarkers for the diagnosis of PAH are limited and resent clinical trials[64] have highlighted the need for makers reflective of drug–target interaction and disease progression. Remodeling of the pulmonary vasculature alters the cellular composition of the vessel wall and results in differential miRNA expression patterns.[48] Little is known of the mechanisms regulating cellular localization and release of miRNA; however, miRNA are stable within multiple peripheral blood fractions.[28,32,49] Levels of circulating miRNA may reflect the progressive vascular remodeling of PAH and could therefore be used to aid diagnosis and to monitor disease progression and response to treatment.

Examination of miRNA expression has helped to elucidate novel signaling pathways and integrate the cellular pathology of PAH. Further technological and computational advances offer the potential for a greater understanding of the processes driving PAH and the identification of novel targets for therapeutic manipulation in a condition that is devastating to patients and their families.

## ACKNOWLEDGMENTS

Supported by the British Heart Foundation Senior Basic Science Research Fellowship FS/13/48/30453 (AL) and the Medical Research Council Clinical Research Training Fellowship MR/K002406/1 (AR).

## REFERENCES

1. Macchia A, Marchioli R, Marfisi R, et al. A meta-analysis of trials of pulmonary hypertension: a clinical condition looking for drugs and research methodology. *Am Heart J.* 2007;153(6):1037–1047.
2. Galie N, Manes A, Negro L, Palazzini M, Bacchi-Reggiani ML, Branzi A. A meta-analysis of randomized controlled trials in pulmonary arterial hypertension. *Eur Heart J.* 2008;30(4):394–403.
3. Morrell NW, Adnot S, Archer SL, et al. Cellular and molecular basis of pulmonary arterial hypertension. *J Am Coll Cardiol.* 2009;54(1 suppl):S20–S31.
4. Machado RD. Investigation of second genetic hits at the BMPR2 locus as a modulator of disease progression in familial pulmonary arterial hypertension. *Circulation.* 2005;111(5):607–613.
5. Humbert M, Monti G, Brenot F, et al. Increased interleukin-1 and interleukin-6 serum concentrations in severe primary pulmonary hypertension. *Am J Respir Crit Care Med.* 1995;151(5):1628–1631.
6. Abenhaim L, Humbert M. Pulmonary hypertension related to drugs and toxins. *Curr Opin Cardiol.* 1999;14(5):437–441.
7. International PPH Consortium, Lane KB, Machado RD, et al. Heterozygous germline mutations in BMPR2, encoding a TGF-beta receptor, cause familial primary pulmonary hypertension. *Nat Genet.* 2000;26(1):81–84.

8. Deng Z, Morse JH, Slager SL, et al. Familial primary pulmonary hypertension (gene PPH1) is caused by mutations in the bone morphogenetic protein receptor-II gene. *Am J Hum Genet.* 2000;67(3):737–744.
9. Rosenberg HC, Rabinovitch M. Endothelial injury and vascular reactivity in monocrotaline pulmonary hypertension. *Am J Physiol.* 1988;255(6 Pt. 2):H1484–H1491.
10. Rabinovitch M, Bothwell T, Hayakawa BN, et al. Pulmonary artery endothelial abnormalities in patients with congenital heart defects and pulmonary hypertension. A correlation of light with scanning electron microscopy and transmission electron microscopy. *Lab Invest.* 1986;55(6):632–653.
11. D'Alonzo GE, Barst RJ, Ayres SM, et al. Survival in patients with primary pulmonary hypertension. Results from a national prospective registry. *Ann Intern Med.* 1991; 115(5):343–349.
12. Xu W, Koeck T, Lara AR, et al. Alterations of cellular bioenergetics in pulmonary artery endothelial cells. *Proc Natl Acad Sci USA.* 2007;104(4):1342–1347.
13. Thenappan T, Shah SJ, Rich S, Gomberg-Maitland M. A USA-based registry for pulmonary arterial hypertension: 1982-2006. *Eur Respir J.* 2007;30(6):1103–1110.
14. Thompson K, Rabinovitch M. Exogenous leukocyte and endogenous elastases can mediate mitogenic activity in pulmonary artery smooth muscle cells by release of extracellular-matrix bound basic fibroblast growth factor. *J Cell Physiol.* 1996;166(3): 495–505.
15. Xu W, Kaneko FT, Zheng S, et al. Increased arginase II and decreased NO synthesis in endothelial cells of patients with pulmonary arterial hypertension. *FASEB J.* 2004;18(14):1746–1748.
16. Stacher E, Graham BB, Hunt JM, et al. Modern age pathology of pulmonary arterial hypertension. *Am J Respir Crit Care Med.* 2012;186(3):261–272.
17. Stenmark KR, Meyrick B, Galie N, Mooi WJ, McMurtry IF. Animal models of pulmonary arterial hypertension: the hope for etiological discovery and pharmacological cure. *Am J Physiol Lung Cell Mol Physiol.* 2009;297(6):L1013–L1032.
18. Taraseviciene-Stewart L, Kasahara Y, Alger L, et al. Inhibition of the VEGF receptor 2 combined with chronic hypoxia causes cell death-dependent pulmonary endothelial cell proliferation and severe pulmonary hypertension. *FASEB J.* 2001;15(2):427–438.
19. Lee RC, Feinbaum RL, Ambros V. The C. elegans heterochronic gene lin-4 encodes small RNAs with antisense complementarity to lin-14. *Cell.* 1993;75(5):843–854.
20. Wightman B, Ha I, Ruvkun G. Posttranscriptional regulation of the heterochronic gene lin-14 by lin-4 mediates temporal pattern formation in C. elegans. *Cell.* 1993; 75(5):855–862.
21. Griffiths-Jones S, Grocock RJ, van Dongen S, Bateman A, Enright AJ. miRBase: microRNA sequences, targets and gene nomenclature. *Nucleic Acids Res.* 2006;34(suppl. 1): D140–D144.
22. Miyaki S, Nakasa T, Otsuki S, et al. MicroRNA-140 is expressed in differentiated human articular chondrocytes and modulates interleukin-1 responses. *Arthritis Rheum.* 2009;60(9):2723–2730.
23. Pasquinelli AE. Non-coding RNA: MicroRNAs and their targets: recognition, regulation and an emerging reciprocal relationship. *Nat Publ Group.* 2012;13(4):271–282.
24. Calin GA, Dumitru CD, Shimizu M, et al. Frequent deletions and down-regulation of micro-RNA genes miR15 and miR16 at 13q14 in chronic lymphocytic leukemia. *Proc Natl Acad Sci USA.* 2002;99(24):15524–15529.
25. Rayner KJ, Esau CC, Hussain FN, et al. Inhibition of miR-33a/b in non-human primates raises plasma HDL and lowers VLDL triglycerides. *Nature.* 2011;478(7369):404–407.
26. Rayner KJ, Sheedy FJ, Esau CC, et al. Antagonism of miR-33 in mice promotes reverse cholesterol transport and regression of atherosclerosis. *J Clin Invest.* 2011;121(7): 2921–2931.

27. Hergenreider E, Heydt S, Tréguer K, et al. Atheroprotective communication between endothelial cells and smooth muscle cells through miRNAs. *Nat Cell Biol.* 2012; 14(3):249–256.
28. Fichtlscherer S, De Rosa S, Fox H, et al. Circulating microRNAs in patients with coronary artery disease. *Circ Res.* 2010;107(5):667–684.
29. van Rooij E, Quiat D, Johnson BA, et al. A family of microRNAs encoded by myosin genes governs myosin expression and muscle performance. *Dev Cell.* 2009;17(5):662–673.
30. Montgomery RL, Hullinger TG, Semus HM, et al. Therapeutic inhibition of miR-208a improves cardiac function and survival during heart failure. *Circulation.* 2011;124(14): 1537–1547.
31. Caruso P, Dempsie Y, Stevens HC, et al. A role for miR-145 in pulmonary arterial hypertension: evidence from mouse models and patient samples. *Circ Res.* 2012;111(3): 290–300.
32. Courboulin A, Paulin R, Giguere NJ, et al. Role for miR-204 in human pulmonary arterial hypertension. *J Exp Med.* 2011;208(3):535–548.
33. Kim J, Kang Y, Kojima Y, et al. An endothelial apelin-FGF link mediated by miR-424 and miR-503 is disrupted in pulmonary arterial hypertension. *Nat Med.* 2012; 19(1):74–82.
34. Janssen HLA, Reesink HW, Lawitz EJ, et al. Treatment of HCV infection by targeting microRNA. *N Engl J Med.* 2013;368(18):1685–1694.
35. Esteller M. Non-coding RNAs in human disease. *Nat Publ Group.* 2011;12(12): 861–874.
36. Corcoran DL, Pandit KV, Gordon B, Bhattacharjee A. Features of mammalian micro-RNA promoters emerge from polymerase II chromatin immunoprecipitation data. *PLoS One.* 2009;4(4):e5279.
37. Ozsolak F, Poling LL, Wang Z, et al. Chromatin structure analyses identify miRNA promoters. *Genes Dev.* 2008;22(22):3172–3183.
38. Heale BSE, Keegan LP, McGurk L, et al. Editing independent effects of ADARs on the miRNA/siRNA pathways. *EMBO J.* 2009;28(20):3145–3156.
39. Song G, Wang L. MiR-433 and miR-127 arise from independent overlapping primary transcripts encoded by the miR-433-127 locus. In: Volff J-N, ed. *PLoS One.* 2008;3(10): e3574.
40. Newman MA, Hammond SM. Emerging paradigms of regulated microRNA processing. *Genes Dev.* 2010;24(11):1086–1092.
41. Lewis BP, Burge CB, Bartel DP. Conserved seed pairing, often flanked by adenosines, indicates that thousands of human genes are microRNA targets. *Cell.* 2005; 120(1):15–20.
42. Grimson A, Farh KK-H, Johnston WK, Garrett-Engele P, Lim LP, Bartel DP. Micro-RNA targeting specificity in mammals: determinants beyond seed pairing. *Mol Cell.* 2007;27(1):91–105.
43. Baek D, Villén J, Shin C, Camargo FD, Gygi SP, Bartel DP. The impact of microRNAs on protein output. *Nature.* 2008;455(7209):64–71.
44. Witkos TM, Koscianska E, Krzyzosiak WJ. Practical aspects of microRNA target prediction. *Curr Mol Med.* 2011;11(2):93–109.
45. Boon RA, Vickers KC. Intercellular transport of microRNAs. *Arterioscler Thromb Vasc Biol.* 2013;33(2):186–192.
46. Creemers EE, Tijsen AJ, Pinto YM. Circulating microRNAs: novel biomarkers and extracellular communicators in cardiovascular disease? *Circ Res.* 2012;110(3): 483–495.
47. Caruso P, MacLean MR, Khanin R, et al. Dynamic changes in lung microRNA profiles during the development of pulmonary hypertension due to chronic hypoxia and monocrotaline. *Arterioscler Thromb Vasc Biol.* 2010;30(4):716–723.

48. Bockmeyer CL, Maegel L, Janciauskiene S, et al. Plexiform vasculopathy of severe pulmonary arterial hypertension and microRNA expression. *J Heart Lung Transplant.* 2012,31(7).764–772.
49. Rhodes CJ, Wharton J, Boon RA, et al. Reduced miR-150 is associated with poor survival in pulmonary arterial hypertension. *Am J Respir Crit Care Med.* 2012;187(3):294–302.
50. Gou D, Ramchandran R, Peng X, et al. miR-210 has an antiapoptotic effect in pulmonary artery smooth muscle cells during hypoxia. *Am J Physiol Lung Cell Mol Physiol.* 2012;303(8):L682–L691.
51. Brock M, Trenkmann M, Gay RE, et al. Interleukin-6 modulates the expression of the bone morphogenic protein receptor type II through a novel STAT3-microRNA cluster 17/92 pathway. *Circ Res.* 2009;104(10):1184–1191.
52. Parikh VN, Jin RC, Rabello S, et al. MicroRNA-21 integrates pathogenic signaling to control pulmonary hypertension: results of a network bioinformatics approach. *Circulation.* 2012;125(12):1520–1532.
53. Pullamsetti SS, Doebele C, Fischer A, et al. Inhibition of microRNA-17 improves lung and heart function in experimental pulmonary hypertension. *Am J Respir Crit Care Med.* 2012;185(4):409–419.
54. Yang S, Banerjee S, de Freitas A, et al. miR-21 regulates chronic hypoxia-induced pulmonary vascular remodeling. *Am J Physiol Lung Cell Mol Physiol.* 2012;302(6): L521–L529.
55. Drake KM, Zygmunt D, Mavrakis L, et al. Altered microRNA processing in heritable pulmonary arterial hypertension. *Am J Respir Crit Care Med.* 2011;184(12):1400–1408.
56. Hausler SFM, Keller A, Chandran PA, et al. Whole blood-derived miRNA profiles as potential new tools for ovarian cancer screening. *Br J Cancer.* 2010;103(5):693–700.
57. Chandra SM, Razavi H, Kim J, et al. Disruption of the Apelin-APJ system worsens hypoxia-induced pulmonary hypertension. *Arterioscler Thromb Vasc Biol.* 2011;31(4): 814–820.
58. Alastalo T-P, Li M, de Jesus Perez V, et al. Disruption of PPARγ/β-catenin–mediated regulation of apelin impairs BMP-induced mouse and human pulmonary arterial EC survival. *J Clin Invest.* 2011;121(9):3735–3746.
59. Bonauer A, Dimmeler S. The microRNA-17 92 cluster: still a miRacle? *Cell Cycle.* 2009;8(23):3866–3873.
60. Brock M, Samillan VJ, Trenkmann M, et al. AntagomiR directed against miR-20a restores functional BMPR2 signalling and prevents vascular remodelling in hypoxia-induced pulmonary hypertension. *Eur Heart J.* 2012; http://dx.doi.org/10.1093/eurheartj/ehs060.
61. Davis-Dusenbery BN, Chan MC, Reno KE, et al. Down-regulation of Kruppel-like factor-4 (KLF4) by microRNA-143/145 is critical for modulation of vascular smooth muscle cell phenotype by transforming growth factor- and bone morphogenetic protein 4. *J Biol Chem.* 2011;286(32):28097–28110.
62. Long X, Miano JM. Transforming growth factor-beta1 (TGF-beta1) utilizes distinct pathways for the transcriptional activation of microRNA 143/145 in human coronary artery smooth muscle cells. *J Biol Chem.* 2011;286(34):30119–30129.
63. Cheng Y, Liu X, Yang J, et al. MicroRNA-145, a novel smooth muscle cell phenotypic marker and modulator, controls vascular neointimal lesion formation. *Circ Res.* 2009;105(2):158–166.
64. Hoeper MM, Barst RJ, Bourge RC, et al. Imatinib mesylate as add-on therapy for pulmonary arterial hypertension: results of the randomized IMPRES study. *Circulation.* 2013;127(10):1128–1138.

CHAPTER FOUR

# Zebrafish as a Model of Cardiac Disease

**Robert N. Wilkinson*, Chris Jopling[†,‡,§], Fredericus J.M. van Eeden[¶]**
*Department of Cardiovascular Science, Medical School, University of Sheffield, Sheffield, United Kingdom
[†]CNRS, UMR-5203, Institut de Génomique Fonctionnelle, Département de Physiologie, Labex Ion Channel Science and Therapeutics, Montpellier, France
[‡]INSERM, U661, Montpellier, France
[§]Universités de Montpellier 1&2, UMR-5203, Montpellier, France
[¶]MRC Centre for Biomedical Genetics, Department of Biomedical Science, University of Sheffield, Sheffield, United Kingdom

## Contents



## Abstract

The zebrafish has been rapidly adopted as a model for cardiac development and disease. The transparency of the embryo, its limited requirement for active oxygen delivery, and ease of use in genetic manipulations and chemical exposure have made it a powerful alternative to rodents. Novel technologies like TALEN/CRISPR-mediated genome engineering and advanced imaging methods will only accelerate its use. Here, we give an overview of heart development and function in the fish and highlight a number of areas where it is most actively contributing to the understanding of cardiac development and disease. We also review the current state of research on a feature that we only could wish to be conserved between fish and human; cardiac regeneration.

*Progress in Molecular Biology and Translational Science*, Volume 124
ISSN 1877-1173
http://dx.doi.org/10.1016/B978-0-12-386930-2.00004-5

# 1. INTRODUCTION

Heart disease is a major cause of morbidity and mortality in the human population and is strongly influenced by genetics. At the start of the first large-scale screens for mutations that affect zebrafish embryonic development,[1,2] several groups were interested in exploiting fish genetics for medical research. Mutants that affect heart and vessel development and function were "top targets" in this respect and pointed toward genes that may have relevance in cardiovascular disease.[3,4] Since these initial screens, the accessibility and ease of use of the fish embryo has stimulated the adoption of zebrafish (and medaka) as a model organism and resulted in specialized screens for cardiovascular mutants, for example,[5–7] functional analysis of candidate cardiovascular disease genes, transgenic disease models,[8] and chemical screens.[9–11] Overall, this work has firmly established the zebrafish as a valid model showing good conservation of gene function and cardiovascular biology. In some cases, the zebrafish is possibly superior to mice, for instance, with respect to heart rate and electrophysiology. However, in other cases differences are quite clear, for instance, the occurrence of cardiac hyperplasia in addition to hypertrophy under conditions of cardiac stress. Nevertheless, with some care it is usually possible to extrapolate results from fish to mammals. One important difference to mammalian embryos is that zebrafish embryos can live for several days independently of circulation which simplifies analysis of mutants that compromise cardiovascular function.[12] Another exciting difference between mammals and fish is the fact that fish myocardial cells can reenter the cell cycle and this gives fish hearts strong regenerative capacity. A detailed molecular understanding of this process may suggest ways in which heart function could be stimulated in mammals after infarct and this has become a fertile field of research. This chapter aims to give an overview of the strengths and weaknesses of the fish as a cardiovascular model.

# 2. ZEBRAFISH HEART DEVELOPMENT

## 2.1. Early patterning

Fate mapping has shown that cardiac progenitors are formed from bilateral marginal domains in the late blastula.[13,14] These cells move dorsally and anteriorly as part of the anterior lateral plate mesoderm. This movement is at least partially controlled by Apelin/Angiotensin II receptor-like 1b signaling, and mutant and morpholino (MO) analysis has shown that this movement is essential for proper induction of the bilateral myocardial progenitors.[15–17] This bilateral mesodermal heart field expresses a number of early transcription factors (e.g., *Hand2*[18]; *Gata4, 5, 6*[19–22]; *Tbx5a*[23]; *Nkx2.5/7*[24–26]) and is thought to be induced during early-axis formation by various signals, including bone morphogenetic protein (BMP),[27] nodal[27] fibroblast growth factor (Fgf),[28,29] with a mainly negative input from retinoic acid (RA)[28,30–32] and promoted by Hedgehog (Hh) signaling.[33] Wnt signaling appears to play a more complex role depending on developmental stage.[34–37] Cardiomyocyte differentiation markers like *cardiac myosin, light polypeptide* 7 (*myl*7 a.k.a. *clmc2*) can first be observed in these cells around 13-somite stage[38] (Fig. 4.1A) and may be controlled by the *myocyte enhancer factor (mef) 2c* genes.[39] The early stages of cardiac development are highly conserved between zebrafish and mammals (see Chapter 2).

The bilateral primordia migrate between endoderm and yolk syncytial layer (YSL) and fuse at the midline to form a cardiac disc around 20 h postfertilization (hpf) (Fig. 4.1B–D). The migration depends on both these tissues.[20,40,41] Sphingosine-1-phosphate (S1P) is crucial for this later migration; a likely scenario is that *mix-type homeobox gene 1* (*mxtx1*)[42] expression in the YSL regulates *spinster2* expression, an S1P transporter. This leads to S1P release from the YSL[43–46] which results in Fibronectin deposition by the endoderm,[47] the substrate for the migrating primordia. After fusion, the cardiac disc undergoes a complex elongation,[48] clockwise rotation,[49] and leftward jog[50] to form an asymmetrically positioned primitive heart tube (Fig. 4.1D and E). Again Bmps, in conjunction with Nodal play an important role in directing the latter two processes.[51,52] As a result, the more centrally located ventricular cells come to lie distally in the heart tube.[53] In analogy to the situation in mammals, a second heart field contributes cells to the distal ventricle and outflow tract and is dependent on Hh, Bmp, Fgf, RA, and transforming growth factor beta (Tgfβ) signals, regulating *isl1, tbx1, latent TGFβ-binding protein 3,* and *mef2cb* expression.[54–59]

The endocardium of the embryonic heart is derived from the ventralmost marginal cells in the cardiac field in early blastula embryos, but these cells are not restricted to this lineage.[60] At later stages, it was shown using *kdrl:gfp* transgenic embryos that the endocardial cells originate from lateral plate mesoderm just anterior to the myocardial primordia and move posteriorly to form a sheet medial to the myocardial precursors (Fig. 4.1A–D). The fusing myocardium surrounds a part of this sheet in a complex movement, creating an endocardial tube within the myocardial tube.[61]

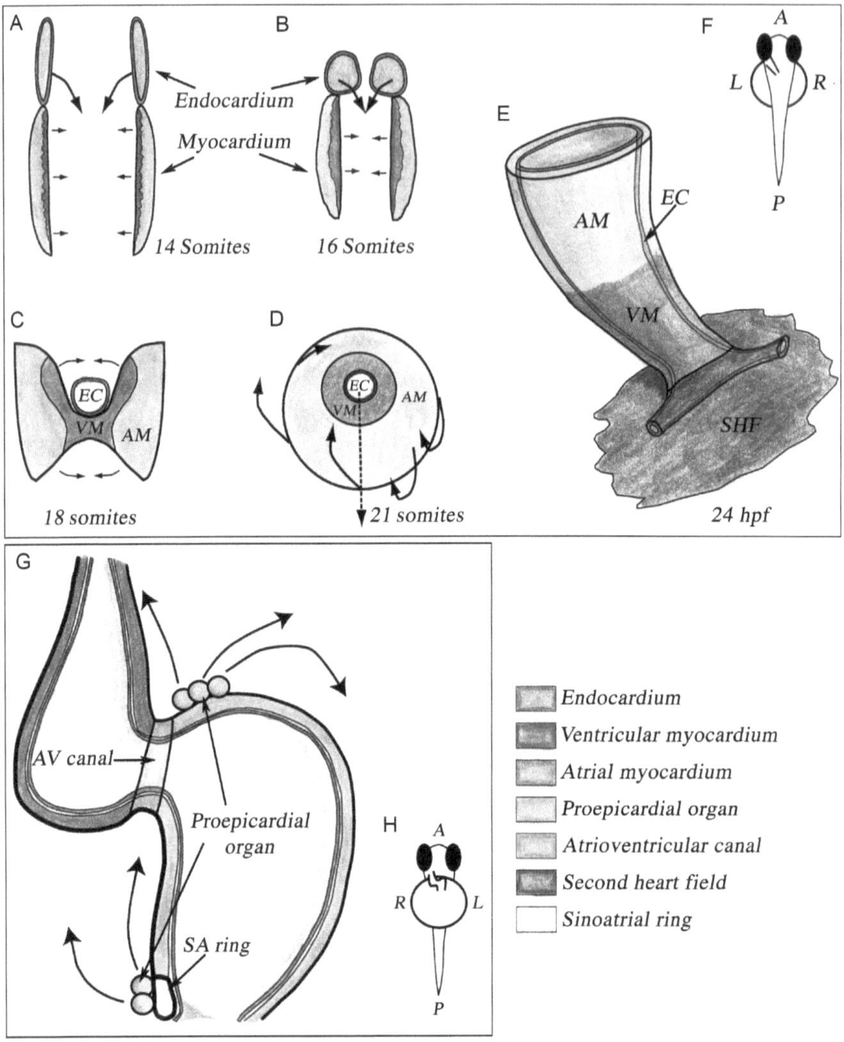


**Figure 4.1** Heart development in zebrafish. (A) At 14-somite stage, bilateral endocardial precursors (EC) are located anterior to myocardial precursors. At 16 somites, the EC precursors move posteriorly, will fuse and come to lie between the myocardial primordia. The bilateral cardiomyocyte (CM) primordia also fuse, first posterior to the endocardial cells (18 somites; C) and then continue to surround the endocardial cells. This forms the cardiac disc (21 somites; D), the most medial myocardial cells will become ventricular cardiomyocytes (VM) whereas more lateral cells are fated to become atrial myocardium (AM). The disc then transforms into a tube (E) by a complex rotation and elongation perpendicular to the plane of the cardiac disc (dashed arrow). The rotation of the disc brings cells on the right side of the disc to the ventral side of the heart tube (i.e., facing the yolk at 24 hpf ), cells on the left will form the "roof" of the heart tube (facing the brain

After initial morphogenesis, the 2 days postfertilization (dpf) zebrafish heart consists of a single atrium and ventricle that are preceded by a sinus venosus and followed by a bulbus arteriosus. Thus, an important contrast to mammals is the single circulation; desaturated venous blood flows into the ventral aorta leading to the gill arches where it is oxygenated and subsequently to the rest of the body. Initially (2 dpf), the heart is subdivided by simple constrictions. The constriction between the atrium and ventricle will become the atrioventricular (AV) canal (AVC) (Fig. 4.1G) that will give rise to the AV valves. The extracellular matrix, also known as cardiac jelly, between endo- and myocardium plays an important role in this process.[62,63] Restricted expression of *versican, bmp4,* and endocardial *notch1b* are early (37–45 hpf) signs of differentiation of the AV region.[64] Transmembrane protein 2, Forkhead box (Fox) n4, and Tbx5a restrict *bmp4* to this domain.[65–67] Foxn4 together with Tbx5 is also required for expression of *tbx2b,* an important regulator of AV canal formation.[66] Protein kinase D2-mediated inactivation of histone deacetylase 5 was recently identified as another essential regulator that restricts *notch1b* to the AVC and is essential for later valve differentiation as shown by expression of Secreted phosphoprotein 1.[68] Similar to mice, Calcineurin-mediated repression of myocardial VEGF expression is also important for valve formation.[6,69,70] Notch1b signaling itself is also involved, but reports on its precise role are contradictory.[6,71] Furthermore, *wnt* signaling is locally activated and required in the AV canal cells for valvulogenesis[72] which also requires a Snail-driven epithelial-mesenchymal transition (EMT).[73] In addition to developmental patterning signals, cardiac function also impacts on valvulogenesis. Mutations and chemicals that block myocardial function will block early endocardial cushion formation involving blood flow-mediated induction of miR-21.[74–76] In addition, retrograde circulatory flows through the AV canal at 2–3 dpf upregulate *notch1b,* possibly via Krüppel-like factor 2a, and this is important for late valve differentiation.[77] This indicates that valve defects can easily arise as a secondary consequence of defective heart

at 24 hpf, E). At 24 hpf, a second heart field (SHF) has been identified which contributes further cells to the heart. (F) Orientation of the fish in (A)–(E): Dorsal view. (G) 48 hpf heart, cross section: atrium and ventricle are connected via the atrioventricular (AV) canal. Cells from the proepicardial organ are formed in two positions. These cells are released in the pericardial space and will move and cover the heart to form the epicardium. At the venous pole of the heart, the sinoatrial (SA) node acts as a pacemaker. (H) Orientation of the fish in (G): ventral view. (See the color plate.)

function and must be taken into consideration, for example, when interpreting valve defects during morphant/mutant analysis.[78]

The heart lies within the pericardial cavity and is covered in epicardium.[79]The epicardium has drawn a lot of attention recently, since it appears to play an important role in cardiac regeneration in fish. It arises from the proepicardial organ which is localized to the AV junction and the sinus venosus at 40 hpf[80] (Fig. 4.1G). These cells are released into the pericardial space due to pericardial fluid motion, will adhere to the heart, and form the epicardium by 4 dpf.[81] Development of the proepicardial organ shows many parallels to its mammalian counterpart and is dependent on the function of Wilms tumor 1 (Wt1), Bmp4, Hand2, Tbx5a, and Suppressor of Ty 6 homolog (Supt6h) function.[80,82]

### 2.2. Larval heart development and growth

At 2 dpf, the heart shows constrictions delineating sinus venosus, atrium, ventricle, and bulbus arteriosus with a single-cell thick myocardium, except in the ventricle where it is two to three cells wide (Fig. 4.1G). At 5 dpf, the atrium has thickened slightly and the ventricle has developed extensive trabeculation, and pectinate muscles start to appear in the atrium. Later development, including a further cardiac rotation is described in Refs. 79,83. After 2 weeks, valves are seen to mature and become thinner and pectinate muscles have become prominent. Going from 4 weeks post fertilization to adulthood, the atrial myocardium thickens to a two to three cell thick layer and the branched pectinate muscles thicken by hypertrophy. The muscular wall of the adult ventricle is three to four cells thick and cell size is increased. Around 6 weeks postfertilization, new layers of cortical muscle form externally to existing muscle, as a result of expansion of a few progenitor-like cells that originate from trabecular muscle, and these cells subsequently "break" through the existing ventricular wall.[84] The epicardium will give rise to the coronal vasculature of the heart. With respect to the valves, post 6 dpf, elongation and maturation in the zebrafish have been described in detail by Martin *et al.* [85]

## 3. FUNCTION AND ELECTROPHYSIOLOGY OF THE ZEBRAFISH HEART

The zebrafish heart is one of the first functional organs in the embryo and it starts beating around 1 dpf, soon after the two myocardial primordia have completed fusion. Although the zebrafish larvae have a simple heart which lacks many of the anatomical features of the mammalian heart, there remain striking parallels with respect to its function, for instance, a pace maker area was defined[86,87] at the sinoatrial (SA) ring which can be considered to be analogous to the mammalian SA node (Fig. 4.1G). With respect to electrophysiology, it has been suggested that the zebrafish heart may actually be more similar to human hearts than mouse hearts, as the latter are dependent for repolarization on $I_{to}$ and $I_{Kslow}$ currents, which are "encoded" by different channels[88–95] (reviewed in Ref. 96).

Using a $Ca^{2+}$-sensitive GFP transgene, development of the cardiac conduction system was found to consist of several developmental steps.[97] Starting at 20 hpf, conduction occurs linearly from sinus venosus to the outflow tract. By 48 hpf. a significant delay of $Ca^{2+}$ signals is seen at the AV boundary, and by 100 hpf, a fast ventricular conduction system is present, which correlates with development of ventricular trabeculae. Further changes are likely to occur during larval development since, for example, the QRS interval duration shortens as the animal matures.[98,99]

In humans, a myocardial cell exerts a typical response when an action potential occurs; in phase 0, an initial stimulus causes sodium channels ($I_{Na}$) to open and induce a fast depolarization. This can be followed by a phase 1 repolarization though opening of transient $K^+$ channels ($I_{to}$). The plateau, "phase 2" of the depolarization is formed by a balance between opening L-type $Ca^{2+}$ channels ($I_{Ca}$) allowing entry of $Ca^{2+}$ and outward $K^+$ movement through delayed rectifier currents. Inactivation of L-type channels and continued opening of $K^+$ channels cause repolarization currents ($I_{Ks}$ and $I_{Kr}$) in phase 3. Final repolarization in phase 4 occurs through outward $K^+$ currents ($I_{K1}$). The $Ca^{2+}$ that enters the cardiomyocyte through the L-type channel also triggers $Ca^{2+}$ release from the sarcoplasmatic reticulum to initiate contraction.[100]

While zebrafish embryos represent a relatively simple easy assay system, the fact that their hearts are immature leaves open the question of their suitability to model adult human hearts with respect to physiology. Analyzing heart physiology in zebrafish adults is complicated due to their aquatic nature and the relatively small size of the adult zebrafish heart (1–2 mm). However, protocols have become available that allow reliable ECG recordings from live adults[101,102] which, in addition to single-cell myocardial recordings, confirm significant pharmacological and electrophysiological parallels between human and zebrafish,[95,103,104] but also a number of differences. Most importantly, there is an absence of a slow rectifying $K^+$ current ($I_{Ks}$) and no clear evidence of $I_{to}$, and a T-type $Ca^{2+}$current has been observed. In human, the latter are only found in fetal hearts or under pathological conditions.

## 4. DISEASE MODELS 1—CONGENITAL HEART DEFECTS

The molecular basis of many congenital heart defects (CHDs), in particular, nonsyndromic and nonfamilial conditions, is still unknown. With an incidence of up to 14/1000,[105] it is an important cause of mortality in newborn infants. Unfortunately, use of the zebrafish in direct modeling of the most commonly found ventricular septal defects is impossible, due to the single circulation in the fish. Nevertheless, the zebrafish and also medaka have contributed to the identification of candidate genes for CHDs, through mutant identification or functional annotation through morphant analysis.[106–111] The function of genes known to be involved in CHD in humans (*GATA4*, *NKX2.5*, *TBX5*, *TBX1*) have largely conserved functions in the zebrafish, although in some cases, increased number of homologs and/or functional shifts between closely related genes were found.[21,67,112] Patients with primary ciliary dyskinesia have a strongly increased risk of CHD as a result of defects in establishing the left-right asymmetry and thus positioning of the heart (heterotaxy).[113] Here, the strengths of the zebrafish and medaka as models for ciliogenesis have contributed and confirmed several candidate genes.[114–118] Currently, fish are being used more and more as a "test tube" to test sequence variants in candidate CHD genes that have been identified in patient DNA.[119–122] In addition, human genetic approaches often identify chromosome segments containing multiple genes, the ease of MO-knockdown and functional analysis of the respective zebrafish homologs can then help to pinpoint the precise gene involved.

## 5. DISEASE MODELS 2—HYPERTROPHIC AND DILATED CARDIOMYOPATHY

Cardiomyopathies are a leading cause of death in young people.[123] Cardiac hypertrophy is the result of an increase in the size but not number of cardiomyocytes and occurs as a normal result of physiological haemodynamic overload, for example, due to intensive physical exercise. Hypertrophic cardiomyopathy (HCM) is a pathological form of this response and can be due to factors such as hypertension, myocardial infarction, or mutations in sarcomeric proteins.[124] Dilated cardiomyopathy (DCM) is another major type of cardiomyopathy, which is characterized by dilation of the ventricle and systolic defects. Several genes have been identified which can lead to either type of cardiomyopathy, thereby indicating a close relationship between the two defects.

One of the first clear parallels between DCM in fish and humans was uncovered by the cloning of a mutation in the titin gene, which was found to be responsible for a familial form of DCM in humans almost simultaneously.[125,126] Knockout or knockdown of numerous other "cardiomyopathy proteins" such as those involved in sarcomere assembly, function, and attachment lead to either type of myopathy and confirm the validity of this fish model.[127–148] Some of these genes are novel candidates for human cardiomyopathies, and zebrafish was used for initial functional characterization,[127,131,132,144,149] whereas other candidates have now already been linked to human cardiomyopathies.[145–147] A further example of the use of fish as a cardiomyopathy model is the functional confirmation or annotation of causative genes like *BAG3, PRDM16,* and *MIB1* following their identification by genetic linkage and sequencing studies.[150–152] Pathological cardiac hypertrophy will lead to fibrosis, disarray of myofibrils, transcriptional activation of fetal cardiac genes, and decreased cardiac output.[153] A robust marker for such cardiomyocyte stress in mammals is the cardiac natriuretic peptide axis, and importantly, the cardiac natriuretic peptide genes (*nppa* and *nppb*) could also be induced by stress in the zebrafish embryonic heart.[9] To be able to capture the dynamic expression of this marker, a luciferase *nppb* reporter has been created and validated as an *in vivo* screening tool for chemical modulators of the hypertrophic response.[9]

A *tnnt2* MO has been used to mimic a dominant mutation causing human cardiac HCM, in order to look at parallels between the effects on the zebrafish embryonic heart and the human heart defect in a more detailed manner.[130] To maximize similarity, the MO injection was titrated to produce a 50% reduction in wild-type message. Microarray analysis showed that the classic hypertrophic transcriptional stress response networks are activated in the *tnnt2* morphant. Myocardial cells also displayed sarcomere disarray, a hallmark of HCM. However, in contrast to the adult human heart, the embryonic hearts of *tnnt* morphants developed cardiomyocyte hyperplasia rather than hypertrophy, probably because fish cardiomyocytes are also able to divide in adults after stress or injury.[154,155] Another interesting result was that $Ca^{2+}$ handling was abnormal in the morphant, suggesting that sarcomere mutations may impinge directly on heart rhythm, rather than that their effect occurs as a consequence of scarring, or abnormal cellular architecture.

A perceived or true lack of oxygen can also lead to pathological cardiac remodeling. *Vhl* mutants have strongly activated hypoxia inducible factor signaling and develop DCM but it is currently unclear whether this is due to intrinsic cardiomyocyte HIF signaling (and resulting VEGF signaling[156,157]) or due to "exhaustion" of the organ as a result of the strongly increased cardiac output.[158] An adult model of pathological cardiac remodeling was created in the form of a *tr265/tr265* line, which is anemic due to mutation of the Band3 protein, which is required in red blood cells. The hearts of these fish showed cardiomegaly, which was initially due to hypertrophy but later on also due to hyperplasia. Thus, the increased regenerative/ proliferative capabilities of the fish heart may be an important difference and suggests that caution is required when extrapolating results from fish to mammals.[159] The line was exploited to show that Wnt signaling promotes hyperplasia in fish.[160]

A further cardiomyopathy model has been created by treatment of fish with doxorubicin.[161] Doxorubicin-treated fish appear to lack the hyperplasia that accompanies anemic models and might be closer to the human situation. This model confirmed evidence that inhibition of mTOR may reduce hypertrophy. A *ztor*/+ background which reduces signaling in both Torc1/Torc2 arms of the mTor pathway showed improvement in heart size, function, and long-term survival compared to wild-type fish, when low doses of doxorubicin were used to induce cardiomyopathy, however, acute cardiotoxicity induced by high doses was exaggerated rather than rescued by reduction of Tor function.

## 6. DISEASE MODELS 3—ARRHYTHMIAS

Heart form and function are closely related during development, but also in the fish environmental factors such as hemodynamic flow and cardiac contraction can modify the cellular architecture of the heart.[74,77] In addition, differences between various mutants affecting heart function indicate that certain genes may in fact have dual functions in both sarcomere formation and generation of heart rhythm. An example of such complex defects is *heart adapter protein* where high dose MO injection leads to defects in AVC and endocardial cushion formation, however, low doses were shown to primarily increase cytoplasmic $Ca^{2+}$ during diastole, thereby prolonging duration of the $Ca^{2+}$ transient. Similarly, the *popeye domain containing 2* gene appears to have a dual function, at high doses trunk muscle fibers and myotendinous junctions were defective, in addition to heart malformations.

However, at lower doses an AV block was observed pointing to defects in the conduction system.[162] Defects in electrical conduction can lead to aberrant patterning of the heart independent of its effect on cardiac contraction; for instance, *cx46* mutants show defects in cellular morphology even in the absence of contractions.[163]

Several channel mutants have been isolated that may model human arrhythmias, including Long QT syndrome. The first heart rhythm mutant to be described was *slow mo,* which has a defect in $I_h$, a pacemaker current,[164,165] and leads to bradycardia. Although cloning of this mutation has not been reported, hyperpolarization-activated cyclic nucleotide-gated channels are good candidates.[166] Depolarization of cardiomyocytes is initiated by sodium currents mainly through SCN5A, in zebrafish two homologs *scn5laa* and *scn5lab,* encoding $Na_v1.5$, have been analyzed. Mutation of this gene in mice leads to death and severe ventricular defects, but analysis is complicated by the strong dependence of the embryo on functional circulation. Also here, a surprising dual role of SCN5 was uncovered, upregulating early *nkx2.5*, *gata4*, and *hand2* expression, but curiously, this role is independent of $I_{Na}$, since chemical modulators of sodium channels failed to reproduce the phenotype.[167–169] A caveat with this study was that mRNA rescue of the MO phenotype was not possible. Interestingly, *nkx2.5* expression is also dependent on another type of channel; a CONNEXIN31.9 orthologue.[170] A human arrhythmogenic SCN5A variant also caused functional defects in fish showing that fish might be used to test novel arrhythmia variants of this and other channels.[171]

High-voltage-activated L-type calcium channels contribute to maintenance of the plateau phase after depolarization. The *island beat* mutant inactivates the α-subunit of L-type calcium channels (Cacna1c).[172] In this mutant, sarcomere development appears unaltered on the ultrastructural level, thus sarcomere abnormalities are not an inevitable consequence of functional deficiencies of the heart or calcium homeostasis. Although this protein does not severely affect sarcomere formation, it does reduce ventricle growth and thickening. Similar effects on growth were found after inactivation of Cacnβ2.1, the regulatory subunit of this channel.[173] This is probably a consequence of abnormal $Ca^{2+}$ homeostasis rather than the lack of contraction, since other contraction mutants such as *tell tale heart* which lacks cardiac myosin light chain-2, do not show such defects.[139,172] In contrast to adult human cardiomyocytes, T-type $Ca^{2+}$ channels are also important in zebrafish hearts.[95] Interestingly, noncanonical *wnt11* signaling directly impinges on the $Ca^{2+}$ homeostasis in order to form the correct electrical gradient during depolarization. Wnt11 appears to shape this gradient by downregulating $Ca^{2+}$ flux through L-type calcium channels, and this process is independent of the planar polarity pathway. Curiously, both *cacna1c* and *wnt* mutants appear to increase intracellular $Ca^{2+}$, which was attributed to increased release from sarcoplasmatic reticulum stores.[99] Another unusual regulator of $Ca^{2+}$ appears to be the Vegf-Flt1-Plcγ1 signaling pathway,[157] which is controlled by Ilk/Pkb/stretch sensing[145,174] but its precise molecular effect remains unknown.

Calcium entry into cardiomyocytes mainly through L-type channels,[175] and further $Ca^{2+}$ release as a result of activation of ryanodine receptors (and possibly Polycystin 2[176]) in the sarcoplasmatic reticulum, initiates contraction. This $Ca^{2+}$ needs to be removed from the cytoplasm to prepare for the next contraction either by sequestration in the sarcoplasmic reticulum via $Ca^{2+}$-ATPase2 or by extrusion into the extracellular space via a cardiac $Na^{+}/Ca^{2+}$ exchanger (Ncx1h), which is mutated in *tremblor* mutants.[177] In this mutant, a defect in calcium handling also appears to result in defective sarcomere assembly.[178] Calcium-release-activated $Ca^{2+}$ channel protein 1 (Orai1) acts to replenish intracellular $Ca^{2+}$ stores in preparation for subsequent $Ca^{2+}$-dependent signaling cascades. In mice, its study is complicated by the strong dependence of the embryo on blood circulation for survival. Orai1 was shown by MO-knockdown to be required for maintenance of sarcomere integrity and z-disk function, partly through regulation of the Calcineurin–Calsarcin–Nfat pathway.[179]

For repolarization of the zebrafish cardiomyocyte, $I_{Kr}$ is important[95] and a functional orthologue of *human Ether-à-go-go-Related Gene* (HERG/KCNH2) is the channel responsible for this. Interestingly, this gene is not the true sequence orthologue of HERG, but *zerg-3/KCNH6*.[180] This gene was found to be mutated in the *breakdance* mutant[3,92] and subsequently, *reggae* mutants were shown to be gain-of-function mutations in this channel.[90] Acquired and genetic forms of long QT syndrome are a clinically important form of arrhythmia and an important risk factor for sudden cardiac death.[181] It has also proven to be an important stumbling block for drug development and use.[182] Thus, there is serious interest in exploiting the zebrafish not only to understand genes and processes involved in QT modulation but also to develop it into a tool to screen drug candidates, either to flag drug leads as being at risk for modulating the QT interval or to find drugs that might "correct" QT abnormalities. Several groups have evaluated zebrafish as a potential assay system and found strong correlation between drug effects on QT interval in patients and induction of bradycardia in fish.[89,92,183–186]

While most genetic mutations leading to QT abnormalities exist in channel proteins, increasingly, other proteins have been implicated. Zebrafish MO knockdowns have provided *in vivo* functional annotation to genes previously not known to be involved in the generation of a normal cardiac cycle[162] and these point to further candidate risk factor genes. In addition, MO/mutant studies can assist genome-wide association studies (GWAS) in identifying and characterizing QT-modulating genes from candidate lists, such a scenario is exemplified by the confirmation of NOS1AP as a modulator of QT length and the identification of GINS3 as a candidate in both fish and GWAS studies.[91]

## 7. HEART REGENERATION

Zebrafish have unique regenerative capabilities, which, in contrast to adult mammals, include the ventricular[187] and atrial myocardium.[8] Adult mammals only show minimal cell division in cardiomyocytes and rather form scar tissue after injury.[188] Several labs are now trying to identify the molecular signals, cells, and tissues that direct cardiac regeneration. A distant goal would be to learn to replicate this process in mammals; more realistically such studies may suggest signaling pathways that could be modulated to improve cellular therapies or endogenous repair capacity. A long list of genes and signaling pathways has been identified in regeneration and are highlighted below and in Table 4.1. There appear to be numerous molecular parallels between fin, brain, and heart regeneration, thus analysis of other regenerative processes in fish may provide further clues.

Resection,[202] cryoinjury,[206–208] or genetic ablation[8] of part of the ventricle in zebrafish lead to rapid activation of various genes in the entire epicardium. For instance, *raldh2* an enzyme required for RA production, is already activated in the epicardium 24 h post injury.[202] In the case of resection/cryoinjury, the wound becomes covered rapidly by the activated epicardium which proliferates and thickens.[8,202] The expression of injury–response genes[32,202,207,208] is then restricted to the area where regeneration occurs. Cells from the epicardium are thought to undergo EMT and express *snail* and *twist* genes.[204]

In the myocardium, following cryoinjury, (which might recapitulate human heart infarcts better than resection) cell death at the site of injury triggers an essential inflammatory response[189] and similarly to other wounding/regeneration assays, leads to production of $H_2O_2$. This promotes tissue infiltration by fibroblasts and leukocytes and raises the possibility that resolution

**Table 4.1** Genes and/or pathways important for regeneration

| Gene/pathway | Likely target | Phenotype | References |
|---|---|---|---|
| Inflammation/ glucocorticoids | Phagocytes | Reduced regeneration after glucocorticoid treatment | 189 |
| Igf/Igf2 | CM | Loss in proliferation and invasion of wound, initial activation (Gata4) normal | 190,191 |
| Gata4 | CM | Scarring instead or regeneration after expression of DN Gata4 | 192 |
| Hedgehog | CM | Increased proliferation after activation by SAG treatment, decreased after inhibition by cyclopamine | 191 |
| Hypoxia/Hif | CM | DN Hif and hyperoxia inhibit regeneration, hypoxia stimulates CM dedifferentiation | 193 |
| Jak/Stat | CM | DN Stat3 blocks regeneration, may act partly via relaxin3a | 193,194 |
| Retinoic acid | CM | *Raldh2* in epi/endocard activated on injury, promotes proliferation of CM | 195 |
| p38α Mapk | CM | Downregulation of p38α Mapk si required but not sufficient for CM proliferation | 196 |
| Cxcl12/ Cxcr4b | CM | Cxcl12 ligand is expressed in epicard, Cxcr4b inhibition (FC131) blocks migration, not proliferation of CMs | 197 |
| Tgfβ | Fibroblasts/ CM | Chemical blockade of Tgfβ (SB431542) blocks regeneration, loss of collagen matrix epi/endocard responses normal, reduced proliferation/ migration of CM | 198,191 |
| Hsp60 | CM | Specifically expressed in blastema, ts-allele blocks regeneration at restrictive temperature | 199 |
| Polo-like kinase 1 | CM | Inhibition of Plk1 with cyclapolin9 inhibits regeneration | 200 |
| *miR133*/Mps1 | CM | *miR133* inhibits proliferation partly via downregulation of Mps1 | 201 |

**Table 4.1** Genes and/or pathways important for regeneration—cont'd

| Gene/pathway | Likely target | Phenotype | References |
|---|---|---|---|
| Fgf17b/ Fgfr2/4 | Epicard | Fgf expression in regenerating CMs induce invasion of epicardial cells, required to re-establish coronal vasculature | 202,203 |
| Pdgf | Epicard | Pdgf is required for epicardial proliferation and coronal vessel formation | 204,205 |

The target cell and a short description of its effects are given.

of inflammation might also occur under the control of $H_2O_2$ as well as via Thymosinβ4-sulfoxide.[209] Later, the fibroblasts deposit a collagen-rich matrix. While this process is occurring, proliferation and dedifferentiation of cortical cardiomyocytes (CMs) occurs at the boundary with the injured area and $Gata4^+$ CMs begin replacing the scar tissue. In embryos, even atrial CMs can contribute to ventricular regeneration.[210] The inability of the CMs to proliferate, for example, due to loss of Igf2[190] leads to a failure of regeneration and permanent scarring, similar to the situation in mammals. Unsurprisingly, cell-cycle regulators such as Polo-like kinase[200] and Mps1 mitotic checkpoint kinase[187] are important in regeneration. *MiR133* is an upstream regulator of *mps1*, and its expression in CMs is reduced during regeneration. Importantly, during regeneration, proliferation is stimulated by downregulation of *miR133*, while proliferation is impaired by elevated *miR133* levels; however, *miR133* has no effect on uninjured hearts. *cx43* was identified in addition to *mps1*, as a likely downstream effector of *miR133*.[201] Although a progenitor cell population was initially thought to be involved,[202] *cre–lox* labeling technology proved that regenerating cells derive from dedifferentiated cardiomyocytes.[200,211] In contrast to mammals, epicardial cells do not contribute CMs. CMs reexpress early cardiac genes like *gata4*,[192,211] *nkx2.5*, *hand2*, *tbx5*, and *tbx20*[202] *msxB/C, and Notch pathway members*.[212,213] Several studies[205,213] have identified numerous additional genes and proteins that are modulated during regeneration, including the homolog of the intriguing *newt Anterior Gradient*[214] gene, various channels, and Jak/Stat signaling components.[193,194] Redifferentiating myocardial cells replace temporary scar tissue and full regeneration occurs within about 60 days.[187] Endocardial cells also proliferate in response to injury as judged from *fli1a:EGFP* transgenic fish[208] and start expressing, for example, *raldh2* and *heart of glass*.[195]

Many signaling pathways play a critical role in the regenerative process. *Fgf17b* is expressed in the regenerating myocardium and may induce delamination of *fgfr2/fgfr4*-positive epicardial cells. These cells contribute to reestablishment of the coronal vasculature[202] but not detectably to myocardial lineages.[203] Pdgf signaling is also essential in the epicardium. Loss of Pdgf leads to defects in epicardial proliferation possibly via Pi3k and defects in delamination, which would normally require Rho-associated protein kinase. Consequently, the coronary vessels fail to reform.[204,205]

Upon injury, *raldh2* is expressed both in epi- and endocardium and leads to production of RA from those cells. Interference with RA signaling inhibits CM proliferation upon injury but has little effect on normal CM proliferation.[195] Tgfβ/Activin signaling is required for regeneration and is activated within the injured area and apposing cardiomyocytes as shown by pSmad3 staining. Blocking the Tgfβ/Activin pathway impairs collagen matrix deposition in the injured area. This is not a trivial consequence of regenerative failure, Fgf inhibition also leads to a block in regeneration, but collagen deposition in the injured area is normal. In addition, absence of this signal impairs proliferation and invasion of myocardial cells into the scar tissue, leading to the suggestion that scar deposition may be beneficial for proper regeneration.[198] Ventricular amputation is likely to lead to hypoxia in the myocardium; interestingly, this is a strong, Hif-dependent stimulus for CMs to dedifferentiate and proliferate.[193]

Neonatal mouse embryos regenerate ventricular myocardium in a similar manner to adult zebrafish.[215,216] However, this regenerative capacity is lost in the first week after birth. p38α Mapk activity is low in embryos but increases after birth and may correlate with a loss of proliferative capacity of CMs. Importantly, inactivation of p38α Mapk together with activation of Fgf signaling can stimulate proliferation of cardiomyocytes.[217] In zebrafish, p38α Mapk activity is present in CMs and exogenous activation of p38α Mapk blocks CM proliferation; however, inactivation of p38α Mapk does not induce ectopic cell division. Therefore, inhibition is a prerequisite but not the sole driving force for proliferation.[196]

## 8. CONCLUDING REMARKS

This review has summarized where the strengths of the zebrafish system lie with respect to cardiac disease. The use of fish to build disease models, which can then be used to not only test and screen for drugs but also identify or characterize disease-causing variants, is of particular significance. Cardiac regeneration may provide vital clues to help repair cardiac damage in mammals and the zebrafish provides a crucial resource to study this process. The creation of reliable fish models of cardiac disease will be speeded up by novel targeted nuclease strategies in addition to the zebrafish mutation project.[218] We expect that the quality of such models will be further improved by application of novel functional fluorescent proteins and cre–lox technology, which are being rapidly adopted by the zebrafish community.

## ACKNOWLEDGMENTS

Supported by EC-FP7 HEALTH-F4-2010-242048 (FvE), a JG Graves Medical Research Fellowship and the Royal Society (RG120564) (RNW), The CDBG is supported by MRC Centre Grant G0700091. C. J. was supported by an INSERM ATIP-AVENIR grant and a Marie Curie C.I.G (CardioGen), C. J. is a member of the Laboratory of Excellence "Ion Channel Science and Therapeutics" supported by a grant from ANR.

## REFERENCES

1. Haffter P, Granato M, Brand M, et al. The identification of genes with unique and essential functions in the development of the zebrafish, Danio rerio. *Development.* 1996;123:1–36.
2. Driever W, Solnica-Krezel L, Schier AF, et al. A genetic screen for mutations affecting embryogenesis in zebrafish. *Development.* 1996;123:37–46.
3. Chen JN, Haffter P, Odenthal J, et al. Mutations affecting the cardiovascular system and other internal organs in zebrafish. *Development.* 1996;123:293–302.
4. Stainier DY, Fouquet B, Chen JN, et al. Mutations affecting the formation and function of the cardiovascular system in the zebrafish embryo. *Development.* 1996;123:285–292.
5. Alexander J, Stainier DY, Yelon D. Screening mosaic F1 females for mutations affecting zebrafish heart induction and patterning. *Dev Genet.* 1998;22:288–299.
6. Beis D, Bartman T, Jin SW, et al. Genetic and cellular analyses of zebrafish atrioventricular cushion and valve development. *Development.* 2005;132:4193–4204.
7. Habeck H, Odenthal J, Walderich B, Maischein H, Schulte-Merker S. Analysis of a zebrafish VEGF receptor mutant reveals specific disruption of angiogenesis. *Curr Biol.* 2002;12:1405–1412.
8. Wang J, Panakova D, Kikuchi K, et al. The regenerative capacity of zebrafish reverses cardiac failure caused by genetic cardiomyocyte depletion. *Development.* 2011;138: 3421–3430.
9. Becker JR, Robinson TY, Sachidanandan C, et al. In vivo natriuretic peptide reporter assay identifies chemical modifiers of hypertrophic cardiomyopathy signalling. *Cardiovasc Res.* 2012;93:463–470.
10. Peal DS, Mills RW, Lynch SN, et al. Novel chemical suppressors of long QT syndrome identified by an in vivo functional screen. *Circulation.* 2010;123:23–30.
11. Hong CC, Peterson QP, Hong JY, Peterson RT. Artery/vein specification is governed by opposing phosphatidylinositol-3 kinase and MAP kinase/ERK signaling. *Curr Biol.* 2006;16:1366–1372.

12. Pelster B, Burggren WW. Disruption of hemoglobin oxygen transport does not impact oxygen-dependent physiological processes in developing embryos of zebra fish (Danio rerio). *Circ Res*. 1996;79:358–362.
13. Stainier DY, Lee RK, Fishman MC. Cardiovascular development in the zebrafish. I. Myocardial fate map and heart tube formation. *Development*. 1993;119:31–40.
14. Keegan BR, Meyer D, Yelon D. Organization of cardiac chamber progenitors in the zebrafish blastula. *Development*. 2004;131:3081–3091.
15. Scott IC, Masri B, D'Amico LA, et al. The g protein-coupled receptor agtrl1b regulates early development of myocardial progenitors. *Dev Cell*. 2007;12:403–413.
16. Zeng XX, Wilm TP, Sepich DS, Solnica-Krezel L. Apelin and its receptor control heart field formation during zebrafish gastrulation. *Dev Cell*. 2007;12:391–402.
17. Paskaradevan S, Scott IC. The Aplnr GPCR regulates myocardial progenitor development via a novel cell-non-autonomous, Galpha(i/o) protein-independent pathway. *Biol Open*. 2012;1:275–285.
18. Yelon D, Ticho B, Halpern ME, et al. The bHLH transcription factor hand2 plays parallel roles in zebrafish heart and pectoral fin development. *Development*. 2000; 127:2573–2582.
19. Serbedzija GN, Chen JN, Fishman MC. Regulation in the heart field of zebrafish. *Development*. 1998;125:1095–1101.
20. Reiter JF, Alexander J, Rodaway A, et al. Gata5 is required for the development of the heart and endoderm in zebrafish. *Genes Dev*. 1999;13:2983–2995.
21. Peterkin T, Gibson A, Patient R. Redundancy and evolution of GATA factor requirements in development of the myocardium. *Dev Biol*. 2007;311:623–635.
22. Holtzinger A, Evans T. Gata5 and Gata6 are functionally redundant in zebrafish for specification of cardiomyocytes. *Dev Biol*. 2007;312:613–622.
23. Begemann G, Ingham PW. Developmental regulation of Tbx5 in zebrafish embryogenesis. *Mech Dev*. 2000;90:299–304.
24. Chen JN, Fishman MC. Zebrafish tinman homolog demarcates the heart field and initiates myocardial differentiation. *Development*. 1996;122:3809–3816.
25. Targoff KL, Colombo S, George V, et al. Nkx genes are essential for maintenance of ventricular identity. *Development*. 2013;140:4203–4213.
26. Lee KH, Xu Q, Breitbart RE. A new tinman-related gene, nkx2.7, anticipates the expression of nkx2.5 and nkx2.3 in zebrafish heart and pharyngeal endoderm. *Dev Biol*. 1996;180:722–731.
27. Reiter JF, Verkade H, Stainier DY. Bmp2b and Oep promote early myocardial differentiation through their regulation of gata5. *Dev Biol*. 2001;234:330–338.
28. Sorrell MR, Waxman JS. Restraint of Fgf8 signaling by retinoic acid signaling is required for proper heart and forelimb formation. *Dev Biol*. 2011;358:44–55.
29. Reifers F, Walsh EC, Leger S, Stainier DY, Brand M. Induction and differentiation of the zebrafish heart requires fibroblast growth factor 8 (fgf8/acerebellar). *Development*. 2000;127:225–235.
30. Keegan BR, Feldman JL, Begemann G, Ingham PW, Yelon D. Retinoic acid signaling restricts the cardiac progenitor pool. *Science*. 2005;307:247–249.
31. Waxman JS, Keegan BR, Roberts RW, Poss KD, Yelon D. Hoxb5b acts downstream of retinoic acid signaling in the forelimb field to restrict heart field potential in zebrafish. *Dev Cell*. 2008;15:923–934.
32. D'Aniello E, Rydeen AB, Anderson JL, Mandal A, Waxman JS. Depletion of retinoic acid receptors initiates a novel positive feedback mechanism that promotes teratogenic increases in retinoic acid. *PLoS Genet*. 2013;9:e1003689.
33. Thomas NA, Koudijs M, van Eeden FJ, Joyner AL, Yelon D. Hedgehog signaling plays a cell-autonomous role in maximizing cardiac developmental potential. *Development*. 2008;135:3789–3799.

34. Ueno S, Weidinger G, Osugi T, et al. Biphasic role for Wnt/beta-catenin signaling in cardiac specification in zebrafish and embryonic stem cells. *Proc Natl Acad Sci USA*. 2007;104:9685–9690.
35. Dohn TE, Waxman JS. Distinct phases of Wnt/beta-catenin signaling direct cardiomyocyte formation in zebrafish. *Dev Biol*. 2012;361:364–376.
36. Rottbauer W, Saurin AJ, Lickert H, et al. Reptin and pontin antagonistically regulate heart growth in zebrafish embryos. *Cell*. 2002;111:661–672.
37. Sorrell MR, Dohn TE, D'Aniello E, Waxman JS. Tcf7l1 proteins cell autonomously restrict cardiomyocyte and promote endothelial specification in zebrafish. *Dev Biol*. 2013;380:199–210.
38. Yelon D, Horne SA, Stainier DY. Restricted expression of cardiac myosin genes reveals regulated aspects of heart tube assembly in zebrafish. *Dev Biol*. 1999;214:23–37.
39. Hinits Y, Hughes SM. Mef2s are required for thick filament formation in nascent muscle fibres. *Development*. 2007;134:2511–2519.
40. Alexander J, Rothenberg M, Henry GL, Stainier DY. Casanova plays an early and essential role in endoderm formation in zebrafish. *Dev Biol*. 1999;215:343–357.
41. Kikuchi Y, Trinh LA, Reiter JF, Alexander J, Yelon D, Stainier DY. The zebrafish bonnie and clyde gene encodes a Mix family homeodomain protein that regulates the generation of endodermal precursors. *Genes Dev*. 2000;14:1279–1289.
42. Sakaguchi T, Kikuchi Y, Kuroiwa A, Takeda H, Stainier DY. The yolk syncytial layer regulates myocardial migration by influencing extracellular matrix assembly in zebrafish. *Development*. 2006;133:4063–4072.
43. Kupperman E, An S, Osborne N, Waldron S, Stainier DY. A sphingosine-1-phosphate receptor regulates cell migration during vertebrate heart development. *Nature*. 2000; 406:192–195.
44. Osborne N, Brand-Arzamendi K, Ober EA, et al. The spinster homolog, two of hearts, is required for sphingosine 1-phosphate signaling in zebrafish. *Curr Biol*. 2008;18: 1882–1888.
45. Kawahara A, Nishi T, Hisano Y, Fukui H, Yamaguchi A, Mochizuki N. The sphingolipid transporter spns2 functions in migration of zebrafish myocardial precursors. *Science*. 2009;323:524–527.
46. Ye D, Lin F. S1pr2/Galpha13 signaling controls myocardial migration by regulating endoderm convergence. *Development*. 2013;140:789–799.
47. Trinh LA, Stainier DY. Fibronectin regulates epithelial organization during myocardial migration in zebrafish. *Dev Cell*. 2004;6:371–382.
48. Horne-Badovinac S, Lin D, Waldron S, et al. Positional cloning of heart and soul reveals multiple roles for PKC lambda in zebrafish organogenesis. *Curr Biol*. 2001; 11:1492–1502.
49. Smith KA, Chocron S, von der Hardt S, et al. Rotation and asymmetric development of the zebrafish heart requires directed migration of cardiac progenitor cells. *Dev Cell*. 2008;14:287–297.
50. Chen JN, van Eeden FJ, Warren KS, et al. Left-right pattern of cardiac BMP4 may drive asymmetry of the heart in zebrafish. *Development*. 1997;124:4373–4382.
51. Veerkamp J, Rudolph F, Cseresnyes Z, et al. Unilateral dampening of Bmp activity by nodal generates cardiac left-right asymmetry. *Dev Cell*. 2013;24:660–667.
52. Lenhart KF, Holtzman NG, Williams JR, Burdine RD. Integration of nodal and BMP signals in the heart requires FoxH1 to create left-right differences in cell migration rates that direct cardiac asymmetry. *PLoS Genet*. 2013;9:e1003109.
53. Glickman NS, Yelon D. Cardiac development in zebrafish: coordination of form and function. *Semin Cell Dev Biol*. 2002;13:507–513.
54. Zhou Y, Cashman TJ, Nevis KR, et al. Latent TGF-beta binding protein 3 identifies a second heart field in zebrafish. *Nature*. 2011;474:645–648.

55. de Pater E, Clijsters L, Marques SR, et al. Distinct phases of cardiomyocyte differentiation regulate growth of the zebrafish heart. *Development*. 2009;136:1633–1641.
56. Hami D, Grimes AC, Tsai HJ, Kirby ML. Zebrafish cardiac development requires a conserved secondary heart field. *Development*. 2011;138:2389–2398.
57. Lazic S, Scott IC. Mef2cb regulates late myocardial cell addition from a second heart field-like population of progenitors in zebrafish. *Dev Biol*. 2011;354:123–133.
58. Guner-Ataman B, Paffett-Lugassy N, Adams MS, et al. Zebrafish second heart field development relies on progenitor specification in anterior lateral plate mesoderm and nkx2.5 function. *Development*. 2013;140:1353–1363.
59. Witzel HR, Jungblut B, Choe CP, Crump JG, Braun T, Dobreva G. The LIM protein Ajuba restricts the second heart field progenitor pool by regulating Isl1 activity. *Dev Cell*. 2012;23:58–70.
60. Lee RK, Stainier DY, Weinstein BM, Fishman MC. Cardiovascular development in the zebrafish. II. Endocardial progenitors are sequestered within the heart field. *Development*. 1994;120:3361–3366.
61. Bussmann J, Bakkers J, Schulte-Merker S. Early endocardial morphogenesis requires Scl/Tal1. *PLoS Genet*. 2007;3:e140.
62. Patra C, Diehl F, Ferrazzi F, et al. Nephronectin regulates atrioventricular canal differentiation via Bmp4-Has2 signaling in zebrafish. *Development*. 2011;138:4499–4509.
63. Peal DS, Burns CG, Macrae CA, Milan D. Chondroitin sulfate expression is required for cardiac atrioventricular canal formation. *Dev Dyn*. 2009;238:3103–3110.
64. Walsh EC, Stainier DY. UDP-glucose dehydrogenase required for cardiac valve formation in zebrafish. *Science*. 2001;293:1670–1673.
65. Smith KA, Lagendijk AK, Courtney AD, et al. Transmembrane protein 2 (Tmem2) is required to regionally restrict atrioventricular canal boundary and endocardial cushion development. *Development*. 2011;138:4193–4198.
66. Chi NC, Shaw RM, De Val S, et al. Foxn4 directly regulates tbx2b expression and atrioventricular canal formation. *Genes Dev*. 2008;22:734–739.
67. Garrity DM, Childs S, Fishman MC. The heartstrings mutation in zebrafish causes heart/fin Tbx5 deficiency syndrome. *Development*. 2002;129:4635–4645.
68. Just S, Berger IM, Meder B, et al. Protein kinase D2 controls cardiac valve formation in zebrafish by regulating histone deacetylase 5 activity. *Circulation*. 2011;124:324–334.
69. Chang CP, Neilson JR, Bayle JH, et al. A field of myocardial-endocardial NFAT signaling underlies heart valve morphogenesis. *Cell*. 2004;118:649–663.
70. Lee YM, Cope JJ, Ackermann GE, et al. Vascular endothelial growth factor receptor signaling is required for cardiac valve formation in zebrafish. *Dev Dyn*. 2006;235:29–37.
71. Timmerman LA, Grego-Bessa J, Raya A, et al. Notch promotes epithelial-mesenchymal transition during cardiac development and oncogenic transformation. *Genes Dev*. 2004;18:99–115.
72. Hurlstone AF, Haramis AP, Wienholds E, et al. The Wnt/beta-catenin pathway regulates cardiac valve formation. *Nature*. 2003;425:633–637.
73. Chen IH, Wang HH, Hsieh YS, Huang WC, Yeh HI, Chuang YJ. PRSS23 is essential for the Snail-dependent endothelial-to-mesenchymal transition during valvulogenesis in zebrafish. *Cardiovasc Res*. 2013;97:443–453.
74. Bartman T, Walsh EC, Wen KK, et al. Early myocardial function affects endocardial cushion development in zebrafish. *PLoS Biol*. 2004;2:E129.
75. Banjo T, Grajcarek J, Yoshino D, et al. Haemodynamically dependent valvulogenesis of zebrafish heart is mediated by flow-dependent expression of miR-21. *Nat Commun*. 2013;4:1978.
76. Kolpa HJ, Peal DS, Lynch SN, et al. miR-21 represses Pdcd4 during cardiac valvulogenesis. *Development*. 2013;140:2172–2180.

77. Vermot J, Forouhar AS, Liebling M, et al. Reversing blood flows act through klf2a to ensure normal valvulogenesis in the developing heart. *PLoS Biol.* 2009;7: e1000246.
78. Glenn NO, McKane M, Kohli V, et al. The W-loop of alpha-cardiac actin is critical for heart function and endocardial cushion morphogenesis in zebrafish. *Mol Cell Biol.* 2012;32:3527–3540.
79. Hu N, Sedmera D, Yost HJ, Clark EB. Structure and function of the developing zebrafish heart. *Anat Rec.* 2000;260:148–157.
80. Serluca FC. Development of the proepicardial organ in the zebrafish. *Dev Biol.* 2008; 315:18–27.
81. Peralta M, Steed E, Harlepp S, et al. Heartbeat-driven pericardiac fluid forces contribute to epicardium morphogenesis. *Curr Biol.* 2013;23:1726–1735.
82. Liu J, Stainier DY. Tbx5 and Bmp signaling are essential for proepicardium specification in zebrafish. *Circ Res.* 2010;106:1818–1828.
83. Singleman C, Holtzman NG. Analysis of postembryonic heart development and maturation in the zebrafish, Danio rerio. *Dev Dyn.* 2012;241:1993–2004.
84. Gupta V, Poss KD. Clonally dominant cardiomyocytes direct heart morphogenesis. *Nature.* 2012;484:479–484.
85. Martin RT, Bartman T. Analysis of heart valve development in larval zebrafish. *Dev Dyn.* 2009;238:1796–1802.
86. Arrenberg AB, Stainier DY, Baier H, Huisken J. Optogenetic control of cardiac function. *Science.* 2010;330:971–974.
87. Tessadori F, van Weerd JH, Burkhard SB, et al. Identification and functional characterization of cardiac pacemaker cells in zebrafish. *PLoS One.* 2012;7:e47644.
88. Arnaout R, Ferrer T, Huisken J, et al. Zebrafish model for human long QT syndrome. *Proc Natl Acad Sci USA.* 2007;104:11316–11321.
89. Milan DJ, Peterson TA, Ruskin JN, Peterson RT, MacRae CA. Drugs that induce repolarization abnormalities cause bradycardia in zebrafish. *Circulation.* 2003;107: 1355–1358.
90. Hassel D, Scholz EP, Trano N, et al. Deficient zebrafish ether-a-go-go-related gene channel gating causes short-QT syndrome in zebrafish reggae mutants. *Circulation.* 2008;117:866–875.
91. Milan DJ, Kim AM, Winterfield JR, et al. Drug-sensitized zebrafish screen identifies multiple genes, including GINS3, as regulators of myocardial repolarization. *Circulation.* 2009;120:553–559.
92. Langheinrich U, Vacun G, Wagner T. Zebrafish embryos express an orthologue of HERG and are sensitive toward a range of QT-prolonging drugs inducing severe arrhythmia. *Toxicol Appl Pharmacol.* 2003;193:370–382.
93. Nerbonne JM. Studying cardiac arrhythmias in the mouse—a reasonable model for probing mechanisms? *Trends Cardiovasc Med.* 2004;14:83–93.
94. Kopp R, Schwerte T, Pelster B. Cardiac performance in the zebrafish breakdance mutant. *J Exp Biol.* 2005;208:2123–2134.
95. Nemtsas P, Wettwer E, Christ T, Weidinger G, Ravens U. Adult zebrafish heart as a model for human heart? An electrophysiological study. *J Mol Cell Cardiol.* 2010;48: 161–171.
96. Leong IU, Skinner JR, Shelling AN, Love DR. Zebrafish as a model for long QT syndrome: the evidence and the means of manipulating zebrafish gene expression. *Acta Physiol (Oxf).* 2010;199:257–276.
97. Chi NC, Shaw RM, Jungblut B, et al. Genetic and physiologic dissection of the vertebrate cardiac conduction system. *PLoS Biol.* 2008;6:e109.
98. Yu F, Huang J, Adlerz K, et al. Evolving cardiac conduction phenotypes in developing zebrafish larvae: implications to drug sensitivity. *Zebrafish.* 2010;7:325–331.

99. Panakova D, Werdich AA, Macrae CA. Wnt11 patterns a myocardial electrical gradient through regulation of the L-type Ca(2+) channel. *Nature*. 2010;466: 874–878.
100. Roden DM, Balser JR, George Jr AL, Anderson ME. Cardiac ion channels. *Annu Rev Physiol*. 2002;64:431–475.
101. Milan DJ, Jones IL, Ellinor PT, MacRae CA. In vivo recording of adult zebrafish electrocardiogram and assessment of drug-induced QT prolongation. *Am J Physiol Heart Circ Physiol*. 2006;291:H269–H273.
102. Sun P, Zhang Y, Yu F, et al. Micro-electrocardiograms to study post-ventricular amputation of zebrafish heart. *Ann Biomed Eng*. 2009;37:890–901.
103. Brette F, Luxan G, Cros C, Dixey H, Wilson C, Shiels HA. Characterization of isolated ventricular myocytes from adult zebrafish (Danio rerio). *Biochem Biophys Res Commun*. 2008;374:143–146.
104. Werdich AA, Brzezinski A, Jeyaraj D, et al. The zebrafish as a novel animal model to study the molecular mechanisms of mechano-electrical feedback in the heart. *Prog Biophys Mol Biol*. 2012;110:154–165.
105. Hoffman JI, Kaplan S. The incidence of congenital heart disease. *J Am Coll Cardiol*. 2002;39:1890–1900.
106. Bamford RN, Roessler E, Burdine RD, et al. Loss-of-function mutations in the EGF-CFC gene CFC1 are associated with human left-right laterality defects. *Nat Genet*. 2000;26:365–369.
107. Szeto DP, Griffin KJ, Kimelman D. HrT is required for cardiovascular development in zebrafish. *Development*. 2002;129:5093–5101.
108. Ahn DG, Ruvinsky I, Oates AC, Silver LM, Ho RK. tbx20, a new vertebrate T-box gene expressed in the cranial motor neurons and developing cardiovascular structures in zebrafish. *Mech Dev*. 2000;95:253–258.
109. Kirk EP, Sunde M, Costa MW, et al. Mutations in cardiac T-box factor gene TBX20 are associated with diverse cardiac pathologies, including defects of septation and valvulogenesis and cardiomyopathy. *Am J Hum Genet*. 2007;81:280–291.
110. Mellman K, Huisken J, Dinsmore C, Hoppe C, Stainier DY. Fibrillin-2b regulates endocardial morphogenesis in zebrafish. *Dev Biol*. 2012;372:111–119.
111. Wang X, Yu Q, Wu Q, et al. Genetic interaction between pku300 and fbn2b controls endocardial cell proliferation and valve development in zebrafish. *J Cell Sci*. 2013;126: 1381–1391.
112. Tu CT, Yang TC, Tsai HJ. Nkx2.7 and Nkx2.5 function redundantly and are required for cardiac morphogenesis of zebrafish embryos. *PLoS One*. 2009;4:e4249.
113. Kennedy MP, Omran H, Leigh MW, et al. Congenital heart disease and other heterotaxic defects in a large cohort of patients with primary ciliary dyskinesia. *Circulation*. 2007;115:2814–2821.
114. Becker-Heck A, Zohn IE, Okabe N, et al. The coiled-coil domain containing protein CCDC40 is essential for motile cilia function and left-right axis formation. *Nat Genet*. 2011;43:79–84.
115. Mitchison HM, Schmidts M, Loges NT, et al. Mutations in axonemal dynein assembly factor DNAAF3 cause primary ciliary dyskinesia. *Nat Genet*. 2012;44:381–389.
116. Loges NT, Olbrich H, Becker-Heck A, et al. Deletions and point mutations of LRRC50 cause primary ciliary dyskinesia due to dynein arm defects. *Am J Hum Genet*. 2009;85:883–889.
117. Omran H, Kobayashi D, Olbrich H, et al. Ktu/PF13 is required for cytoplasmic pre-assembly of axonemal dyneins. *Nature*. 2008;456:611–616.
118. van Rooijen E, Giles RH, Voest EE, van Rooijen C, Schulte-Merker S, van Eeden FJ. LRRC50, a conserved ciliary protein implicated in polycystic kidney disease. *J Am Soc Nephrol*. 2008;19:1128–1138.

119. Kodo K, Nishizawa T, Furutani M, et al. Genetic analysis of essential cardiac transcription factors in 256 patients with non-syndromic congenital heart defects. *Circ J*. 2012;76:1703–1711.
120. Smith KA, Joziasse IC, Chocron S, et al. Dominant-negative ALK2 allele associates with congenital heart defects. *Circulation*. 2009;119:3062–3069.
121. Karkera JD, Lee JS, Roessler E, et al. Loss-of-function mutations in growth differentiation factor-1 (GDF1) are associated with congenital heart defects in humans. *Am J Hum Genet*. 2007;81:987–994.
122. Hyde AS, Farmer EL, Easley KE, et al. UDP-glucose dehydrogenase polymorphisms from patients with congenital heart valve defects disrupt enzyme stability and quaternary assembly. *J Biol Chem*. 2012;287:32708–32716.
123. Maron BJ. Hypertrophic cardiomyopathy: a systematic review. *JAMA*. 2002;287: 1308–1320.
124. Barry SP, Davidson SM, Townsend PA. Molecular regulation of cardiac hypertrophy. *Int J Biochem Cell Biol*. 2008;40:2023–2039.
125. Xu X, Meiler SE, Zhong TP, et al. Cardiomyopathy in zebrafish due to mutation in an alternatively spliced exon of titin. *Nat Genet*. 2002;30:205–209.
126. Gerull B, Gramlich M, Atherton J, et al. Mutations of TTN, encoding the giant muscle filament titin, cause familial dilated cardiomyopathy. *Nat Genet*. 2002;30: 201–204.
127. Poon KL, Tan KT, Wei YY, et al. RNA-binding protein RBM24 is required for sarcomere assembly and heart contractility. *Cardiovasc Res*. 2012;94:418–427.
128. Van't Padje S, Chaudhry B, Severijnen LA, et al. Reduction in fragile X related 1 protein causes cardiomyopathy and muscular dystrophy in zebrafish. *J Exp Biol*. 2009;212: 2564–2570.
129. Sehnert AJ, Huq A, Weinstein BM, Walker C, Fishman M, Stainier DY. Cardiac troponin T is essential in sarcomere assembly and cardiac contractility. *Nat Genet*. 2002; 31:106–110.
130. Becker JR, Deo RC, Werdich AA, Panakova D, Coy S, MacRae CA. Human cardiomyopathy mutations induce myocyte hyperplasia and activate hypertrophic pathways during cardiogenesis in zebrafish. *Dis Model Mech*. 2011;4:400–410.
131. Will RD, Eden M, Just S, et al. Myomasp/LRRC39, a heart- and muscle-specific protein, is a novel component of the sarcomeric M-band and is involved in stretch sensing. *Circ Res*. 2010;107:1253–1264.
132. Seeger TS, Frank D, Rohr C, et al. Myozap, a novel intercalated disc protein, activates serum response factor-dependent signaling and is required to maintain cardiac function in vivo. *Circ Res*. 2010;106:880–890.
133. Vogel B, Meder B, Just S, et al. In-vivo characterization of human dilated cardiomyopathy genes in zebrafish. *Biochem Biophys Res Commun*. 2009;390:516–522.
134. Ho YL, Lin YH, Tsai WY, Hsieh FJ, Tsai HJ. Conditional antisense-knockdown of zebrafish cardiac troponin C as a new animal model for dilated cardiomyopathy. *Circ J*. 2009;73:1691–1697.
135. Meder B, Laufer C, Hassel D, et al. A single serine in the carboxyl terminus of cardiac essential myosin light chain-1 controls cardiomyocyte contractility in vivo. *Circ Res*. 2009;104:650–659.
136. Martin ED, Moriarty MA, Byrnes L, Grealy M. Plakoglobin has both structural and signalling roles in zebrafish development. *Dev Biol*. 2009;327:83–96.
137. Wang L, Sewell WF, Kim SD, et al. Eya4 regulation of Na+/K+-ATPase is required for sensory system development in zebrafish. *Development*. 2008;135:3425–3434.
138. Chen Z, Huang W, Dahme T, Rottbauer W, Ackerman MJ, Xu X. Depletion of zebrafish essential and regulatory myosin light chains reduces cardiac function through distinct mechanisms. *Cardiovasc Res*. 2008;79:97–108.

139. Rottbauer W, Wessels G, Dahme T, et al. Cardiac myosin light chain-2: a novel essential component of thick-myofilament assembly and contractility of the heart. *Circ Res.* 2006;99:323–331.
140. Heuser A, Plovie ER, Ellinor PT, et al. Mutant desmocollin-2 causes arrhythmogenic right ventricular cardiomyopathy. *Am J Hum Genet.* 2006;79:1081–1088.
141. Khuchua Z, Yue Z, Batts L, Strauss AW. A zebrafish model of human Barth syndrome reveals the essential role of tafazzin in cardiac development and function. *Circ Res.* 2006;99:201–208.
142. Schonberger J, Wang L, Shin JT, et al. Mutation in the transcriptional coactivator EYA4 causes dilated cardiomyopathy and sensorineural hearing loss. *Nat Genet.* 2005;37:418–422.
143. Cheng L, Guo XF, Yang XY, et al. Delta-sarcoglycan is necessary for early heart and muscle development in zebrafish. *Biochem Biophys Res Commun.* 2006;344: 1290–1299.
144. Seguchi O, Takashima S, Yamazaki S, et al. A cardiac myosin light chain kinase regulates sarcomere assembly in the vertebrate heart. *J Clin Invest.* 2007;117:2812–2824.
145. Bendig G, Grimmler M, Huttner IG, et al. Integrin-linked kinase, a novel component of the cardiac mechanical stretch sensor, controls contractility in the zebrafish heart. *Genes Dev.* 2006;20:2361–2372.
146. Hassel D, Dahme T, Erdmann J, et al. Nexilin mutations destabilize cardiac Z-disks and lead to dilated cardiomyopathy. *Nat Med.* 2009;15:1281–1288.
147. Knoll R, Postel R, Wang J, et al. Laminin-alpha4 and integrin-linked kinase mutations cause human cardiomyopathy via simultaneous defects in cardiomyocytes and endothelial cells. *Circulation.* 2007;116:515–525.
148. Yang J, Xu X. alpha-Actinin2 is required for the lateral alignment of Z discs and ventricular chamber enlargement during zebrafish cardiogenesis. *FASEB J.* 2012;26: 4230–4242.
149. Spaich S, Will RD, Just S, et al. F-box and leucine-rich repeat protein 22 is a cardiac-enriched F-box protein that regulates sarcomeric protein turnover and is essential for maintenance of contractile function in vivo. *Circ Res.* 2012;111:1504–1516.
150. Norton N, Li D, Rieder MJ, et al. Genome-wide studies of copy number variation and exome sequencing identify rare variants in BAG3 as a cause of dilated cardiomyopathy. *Am J Hum Genet.* 2011;88:273–282.
151. Arndt AK, Schafer S, Drenckhahn JD, et al. Fine mapping of the 1p36 deletion syndrome identifies mutation of PRDM16 as a cause of cardiomyopathy. *Am J Hum Genet.* 2013;93:67–77.
152. Luxan G, Casanova JC, Martinez-Poveda B, et al. Mutations in the NOTCH pathway regulator MIB1 cause left ventricular noncompaction cardiomyopathy. *Nat Med.* 2013;19:193–201.
153. Ahmad F, Seidman JG, Seidman CE. The genetic basis for cardiac remodeling. *Annu Rev Genomics Hum Genet.* 2005;6:185–216.
154. Jean MJ, Deverteuil P, Lopez NH, Tapia JD, Schoffstall B. Adult zebrafish hearts efficiently compensate for excessive forced overload cardiac stress with hyperplastic cardiomegaly. *Biores Open Access.* 2012;1:88–91.
155. Parente V, Balasso S, Pompilio G, et al. Hypoxia/reoxygenation cardiac injury and regeneration in zebrafish adult heart. *PLoS One.* 2013;8:e53748.
156. Kim CH, Cho YS, Chun YS, Park JW, Kim MS. Early expression of myocardial HIF-1alpha in response to mechanical stresses: regulation by stretch-activated channels and the phosphatidylinositol 3-kinase signaling pathway. *Circ Res.* 2002;90: E25–E33.
157. Rottbauer W, Just S, Wessels G, et al. VEGF-PLCgamma1 pathway controls cardiac contractility in the embryonic heart. *Genes Dev.* 2005;19:1624–1634.

158. van Rooijen E, Voest EE, Logister I, et al. Zebrafish mutants in the von Hippel-Lindau tumor suppressor display a hypoxic response and recapitulate key aspects of Chuvash polycythemia. *Blood*. 2009;113:6449–6460.
159. Sun X, Hoage T, Bai P, et al. Cardiac hypertrophy involves both myocyte hypertrophy and hyperplasia in anemic zebrafish. *PLoS One*. 2009;4:e6596.
160. Hoage T, Sun X, Xu X. Functions of the Wnt/beta-catenin pathway in an anemia-induced zebrafish model of cardiomyopathy are location dependent. *Biochem Biophys Res Commun*. 2011;415:490–496.
161. Ding Y, Sun X, Huang W, et al. Haploinsufficiency of target of rapamycin attenuates cardiomyopathies in adult zebrafish. *Circ Res*. 2011;109:658–669.
162. Kirchmaier BC, Poon KL, Schwerte T, et al. The Popeye domain containing 2 (popdc2) gene in zebrafish is required for heart and skeletal muscle development. *Dev Biol*. 2012;363:438–450.
163. Chi NC, Bussen M, Brand-Arzamendi K, et al. Cardiac conduction is required to preserve cardiac chamber morphology. *Proc Natl Acad Sci USA*. 2010;107:14662–14667.
164. Baker K, Warren KS, Yellen G, Fishman MC. Defective "pacemaker" current (Ih) in a zebrafish mutant with a slow heart rate. *Proc Natl Acad Sci USA*. 1997;94:4554–4559.
165. Warren KS, Baker K, Fishman MC. The slow mo mutation reduces pacemaker current and heart rate in adult zebrafish. *Am J Physiol Heart Circ Physiol*. 2001;281: H1711–H1719.
166. Baruscotti M, Bottelli G, Milanesi R, DiFrancesco JC, DiFrancesco D. HCN-related channelopathies. *Pflugers Arch*. 2010;460:405–415.
167. Novak AE, Jost MC, Lu Y, Taylor AD, Zakon HH, Ribera AB. Gene duplications and evolution of vertebrate voltage-gated sodium channels. *J Mol Evol*. 2006;63:208–221.
168. Chopra SS, Stroud DM, Watanabe H, et al. Voltage-gated sodium channels are required for heart development in zebrafish. *Circ Res*. 2010;106:1342–1350.
169. Bennett JS, Stroud DM, Becker JR, Roden DM. Proliferation of embryonic cardiomyocytes in zebrafish requires the sodium channel scn5Lab. *Genesis*. 2013; 51:562–574.
170. Sultana N, Nag K, Hoshijima K, Laird DW, Kawakami A, Hirose S. Zebrafish early cardiac connexin, Cx36.7/Ecx, regulates myofibril orientation and heart morphogenesis by establishing Nkx2.5 expression. *Proc Natl Acad Sci USA*. 2008;105:4763–4768.
171. Huttner IG, Trivedi G, Jacoby A, Mann SA, Vandenberg JI, Fatkin D. A transgenic zebrafish model of a human cardiac sodium channel mutation exhibits bradycardia, conduction-system abnormalities and early death. *J Mol Cell Cardiol*. 2013;61:123–132.
172. Rottbauer W, Baker K, Wo ZG, Mohideen MA, Cantiello HF, Fishman MC. Growth and function of the embryonic heart depend upon the cardiac-specific L-type calcium channel alpha1 subunit. *Dev Cell*. 2001;1:265–275.
173. Chernyavskaya Y, Ebert AM, Milligan E, Garrity DM. The voltage-gated calcium channel CACNB2 (beta2.1) protein is required in the heart for control of cell proliferation and heart tube integrity. *Dev Dyn*. 2012;241:648–662.
174. Meder B, Huttner IG, Sedaghat-Hamedani F, et al. PINCH proteins regulate cardiac contractility by modulating integrin-linked kinase-protein kinase B signaling. *Mol Cell Biol*. 2011;31:3424–3435.
175. Bovo E, Dvornikov AV, Mazurek SR, de Tombe PP, Zima AV. Mechanisms of Ca handling in zebrafish ventricular myocytes. *Pflugers Arch*. 2013;465:1775–1784.
176. Paavola J, Schliffke S, Rossetti S, et al. Polycystin-2 mutations lead to impaired calcium cycling in the heart and predispose to dilated cardiomyopathy. *J Mol Cell Cardiol*. 2013;58:199–208.
177. Langenbacher AD, Dong Y, Shu X, et al. Mutation in sodium-calcium exchanger 1 (NCX1) causes cardiac fibrillation in zebrafish. *Proc Natl Acad Sci USA*. 2005;102: 17699–17704.

178. Ebert AM, Hume GL, Warren KS, et al. Calcium extrusion is critical for cardiac morphogenesis and rhythm in embryonic zebrafish hearts. *Proc Natl Acad Sci USA*. 2005;102:17705–17710.
179. Volkers M, Dolatabadi N, Gude N, Most P, Sussman MA, Hassel D. Orai1 deficiency leads to heart failure and skeletal myopathy in zebrafish. *J Cell Sci*. 2012;125:287–294.
180. Leong IU, Skinner JR, Shelling AN, Love DR. Identification and expression analysis of kcnh2 genes in the zebrafish. *Biochem Biophys Res Commun*. 2010;396:817–824.
181. van Noord C, Eijgelsheim M, Stricker BH. Drug- and non-drug-associated QT interval prolongation. *Br J Clin Pharmacol*. 2010;70:16–23.
182. Roden DM. Drug-induced prolongation of the QT interval. *N Engl J Med*. 2004;350: 1013–1022.
183. Lee SH, Kim HR, Han RX, Oqani RK, Jin DI. Cardiovascular risk assessment of atypical antipsychotic drugs in a zebrafish model. *J Appl Toxicol*. 2013;33: 466–470.
184. Mittelstadt SW, Hemenway CL, Craig MP, Hove JR. Evaluation of zebrafish embryos as a model for assessing inhibition of hERG. *J Pharmacol Toxicol Methods*. 2008;57: 100–105.
185. Dhillon SS, Doro E, Magyary I, Egginton S, Sik A, Muller F. Optimisation of embryonic and larval ECG measurement in zebrafish for quantifying the effect of QT prolonging drugs. *PLoS One*. 2013;8:e60552.
186. Park MJ, Lee KR, Shin DS, et al. Predicted drug-induced bradycardia related cardio toxicity using a zebrafish in vivo model is highly correlated with results from in vitro tests. *Toxicol Lett*. 2013;216:9–15.
187. Poss KD, Wilson LG, Keating MT. Heart regeneration in zebrafish. *Science*. 2002;298:2188–2190.
188. Laflamme MA, Murry CE. Heart regeneration. *Nature*. 2011;473:326–335.
189. Huang WC, Yang CC, Chen IH, Liu YM, Chang SJ, Chuang YJ. Treatment of glucocorticoids inhibited early immune responses and impaired cardiac repair in adult zebrafish. *PLoS One*. 2013;8:e66613.
190. Huang Y, Harrison MR, Osorio A, et al. Igf signaling is required for cardiomyocyte proliferation during zebrafish heart development and regeneration. *PLoS One*. 2013; 8:e67266.
191. Choi WY, Gemberling M, Wang J, et al. In vivo monitoring of cardiomyocyte proliferation to identify chemical modifiers of heart regeneration. *Development*. 2013;140: 660–666.
192. Gupta V, Gemberling M, Karra R, Rosenfeld GE, Evans T, Poss KD. An injury-responsive gata4 program shapes the zebrafish cardiac ventricle. *Curr Biol*. 2013;23: 1221–1227.
193. Jopling C, Sune G, Faucherre A, Fabregat C, Izpisua Belmonte JC. Hypoxia induces myocardial regeneration in zebrafish. *Circulation*. 2012;126:3017–3027.
194. Fang Y, Gupta V, Karra R, Holdway JE, Kikuchi K, Poss KD. Translational profiling of cardiomyocytes identifies an early Jak1/Stat3 injury response required for zebrafish heart regeneration. *Proc Natl Acad Sci USA*. 2013;110:13416–13421.
195. Kikuchi K, Holdway JE, Major RJ, et al. Retinoic acid production by endocardium and epicardium is an injury response essential for zebrafish heart regeneration. *Dev Cell* 2011;20:397–404.
196. Jopling C, Sune G, Morera C, Izpisua Belmonte JC. p38alpha MAPK regulates myocardial regeneration in zebrafish. *Cell Cycle*. 2012;11:1195–1201.
197. Itou J, Oishi I, Kawakami H, et al. Migration of cardiomyocytes is essential for heart regeneration in zebrafish. *Development*. 2012;139:4133–4142.
198. Chablais F, Jazwinska A. The regenerative capacity of the zebrafish heart is dependent on TGFbeta signaling. *Development*. 2012;139:1921–1930.

199. Makino S, Whitehead GG, Lien CL, et al. Heat-shock protein 60 is required for blastema formation and maintenance during regeneration. *Proc Natl Acad Sci USA*. 2005; 102:14599–14604
200. Jopling C, Sleep E, Raya M, Marti M, Raya A, Belmonte JC. Zebrafish heart regeneration occurs by cardiomyocyte dedifferentiation and proliferation. *Nature*. 2010;464: 606–609.
201. Yin VP, Lepilina A, Smith A, Poss KD. Regulation of zebrafish heart regeneration by miR-133. *Dev Biol*. 2012;365:319–327.
202. Lepilina A, Coon AN, Kikuchi K, et al. A dynamic epicardial injury response supports progenitor cell activity during zebrafish heart regeneration. *Cell*. 2006;127:607–619.
203. Kikuchi K, Gupta V, Wang J, et al. tcf21+ epicardial cells adopt non-myocardial fates during zebrafish heart development and regeneration. *Development*. 2011;138: 2895–2902.
204. Kim J, Wu Q, Zhang Y, et al. PDGF signaling is required for epicardial function and blood vessel formation in regenerating zebrafish hearts. *Proc Natl Acad Sci USA*. 2010;107:17206–17210.
205. Lien CL, Schebesta M, Makino S, Weber GJ, Keating MT. Gene expression analysis of zebrafish heart regeneration. *PLoS Biol*. 2006;4:e260.
206. Chablais F, Veit J, Rainer G, Jazwinska A. The zebrafish heart regenerates after cryoinjury-induced myocardial infarction. *BMC Dev Biol*. 2011;11:21.
207. Gonzalez-Rosa JM, Martin V, Peralta M, Torres M, Mercader N. Extensive scar formation and regression during heart regeneration after cryoinjury in zebrafish. *Development*. 2011;138:1663–1674.
208. Schnabel K, Wu CC, Kurth T, Weidinger G. Regeneration of cryoinjury induced necrotic heart lesions in zebrafish is associated with epicardial activation and cardiomyocyte proliferation. *PLoS One*. 2011;6:e18503.
209. Evans MA, Smart N, Dube KN, et al. Thymosin beta4-sulfoxide attenuates inflammatory cell infiltration and promotes cardiac wound healing. *Nat Commun*. 2013;4:2081.
210. Zhang R, Han P, Yang H, et al. In vivo cardiac reprogramming contributes to zebrafish heart regeneration. *Nature*. 2013;498:497–501.
211. Kikuchi K, Holdway JE, Werdich AA, et al. Primary contribution to zebrafish heart regeneration by gata4(+) cardiomyocytes. *Nature*. 2010;464:601–605.
212. Raya A, Koth CM, Buscher D, et al. Activation of Notch signaling pathway precedes heart regeneration in zebrafish. *Proc Natl Acad Sci USA*. 2003;100(Suppl. 1): 11889–11895.
213. Sleep E, Boue S, Jopling C, Raya M, Raya A, Izpisua Belmonte JC. Transcriptomics approach to investigate zebrafish heart regeneration. *J Cardiovasc Med (Hagerstown)*. 2010;11:369–380.
214. Kumar A, Godwin JW, Gates PB, Garza-Garcia AA, Brockes JP. Molecular basis for the nerve dependence of limb regeneration in an adult vertebrate. *Science*. 2007;318: 772–777.
215. Porrello ER, Mahmoud AI, Simpson E, et al. Transient regenerative potential of the neonatal mouse heart. *Science*. 2011;331:1078–1080.
216. Drenckhahn JD, Schwarz QP, Gray S, et al. Compensatory growth of healthy cardiac cells in the presence of diseased cells restores tissue homeostasis during heart development. *Dev Cell*. 2008;15:521–533.
217. Engel FB, Schebesta M, Duong MT, et al. p38 MAP kinase inhibition enables proliferation of adult mammalian cardiomyocytes. *Genes Dev*. 2005;19:1175–1187.
218. Kettleborough RN, Busch-Nentwich EM, Harvey SA, et al. A systematic genome-wide analysis of zebrafish protein-coding gene function. *Nature*. 2013;496:494–497.

CHAPTER FIVE

# The Zebrafish as a Model of Vascular Development and Disease

**Robert N. Wilkinson*, Fredericus J.M. van Eeden†**
*Department of Cardiovascular Science, Medical School, University of Sheffield, Sheffield, United Kingdom
†MRC Centre for Biomedical Genetics, Department of Biomedical Science, University of Sheffield, Sheffield, United Kingdom

## Contents



## Abstract

The zebrafish has recently emerged as an important animal model to study the formation of the vertebrate vascular network. The small size, optical translucency, and genetic tractability of the zebrafish embryo, in combination with an abundance of fluorescent transgenic lines which permit direct visualization of *in vivo* vessel formation, have greatly advanced our understanding of vascular biology. Widespread adoption of this powerful system has led to many important discoveries in relation to the mechanisms that underlie blood vessel formation. This review highlights the contribution of the zebrafish

*Progress in Molecular Biology and Translational Science*, Volume 124
ISSN 1877-1173
http://dx.doi.org/10.1016/B978-0-12-386930-2.00005-7

system to the current understanding of blood vessel formation and the use of zebrafish to model human vascular disease.

## 1. INTRODUCTION

The formation of a correctly patterned and connected network of blood vessels is essential for embryonic survival; hence, the vasculature is one of the first organ systems to arise during vertebrate embryogenesis. The development of blood vessels is a continual process, which not only ensures that the constantly changing physiological requirements of developing embryos are met, but is also integral to many other processes. The vascular tree comprises the network through which immune surveillance occurs,[1] while the embryonic artery is the source of hematopoietic stem cells from which all adult blood lineages are derived.[2] Therefore, defects in the structure or function of blood vessels are responsible for a wide range of congenital and acquired diseases, for example, myocardial infarction, stroke, cerebral cavernous malformations (CCMs), and hereditary hemorrhagic telangiectasia (HHT). The vasculature also plays a critical role in the pathogenesis of diseases such as cancer, since tumors are able to hijack endogenous signaling pathways responsible for regulating the growth and remodeling of blood vessels. Understanding the mechanisms by which blood vessels develop, are patterned and ultimately differentiate, is therefore of fundamental importance to our knowledge of how to combat a diverse range of human diseases.

Throughout evolution, the genetic mechanisms governing endothelial development have been remarkably well conserved between zebrafish and humans. The optical translucency of the zebrafish embryo, *ex utero* development, high fecundity of zebrafish, and an abundance of cardiovascular-specific transgenic lines permit real-time *in vivo* observation of vascular differentiation. Indeed, the use of fluorescent reporter transgenesis in zebrafish embryos allows direct visualization of the expression of any gene of interest, thereby enabling real-time tracking and lineage analysis of individual endothelial cells (ECs) within the developing vasculature. The high level of homology and accompanying genetic tractability have helped establish the zebrafish as a powerful model whose utility has placed it at the forefront of research into vascular development and disease. Furthermore, at the time of writing, 2601 or 82% of the 3176 genes bearing morbidity descriptions in the Online Mendelian Inheritance in Man database can be related to at least one zebrafish ortholog,[3] clearly demonstrating the utility of this model in studying human disease.

## 2. ORIGINS OF THE VASCULATURE IN ZEBRAFISH

The first major vessels of the vertebrate embryo are generated via a process known as vasculogenesis. Vasculogenesis is the mechanism by which free, often migratory vascular ECs, coalesce to form vascular cords *in situ* from which lumenized vessels subsequently develop. Following the vasculogenic formation of primary vessels, the initial vascular network is expanded by the complex secondary process of angiogenesis, which remodels pre-existing vessels either by sprouting and elongation or by the division or fusion of existing vessels. During these processes, vessels become specified as arteries and veins and subsequently recruit supportive smooth muscle cells and pericytes to protect the endothelium and maintain vascular integrity. The precise genetic programming required to shape and pattern the vasculature is still only partly understood, in part due to not only the complexity of the multiple cellular interactions involved in its formation but also the molecular complexity present within those interactions.[4] However, the zebrafish has allowed us to unravel much of this complexity through the discovery and characterization of many different signaling pathways which function during vascular network formation (reviewed in Ref. 5). The utility of the zebrafish system has enabled genetic dissection of the mechanisms by which the vascular system develops and differentiates, and how these processes are dysregulated in disease.

Through the use of cell lineage tracing, it is known that the blood and endothelium derive from ventral mesoderm, the induction and patterning of which is due to the actions of the TGF-β family, namely, Activin/Nodal and Bmp; the Wnt pathway, and the FGF pathway.[6–8] Members of the Bmp family play a fundamental role in dorsoventral patterning, and an activity gradient of these secreted molecules is instrumental in inducing ventral fate in developing embryos.[9–13] High levels of Bmp signaling promote the formation of ventral mesoderm, while lower levels of Bmp activity, arising from direct Bmp antagonism, promote the formation of dorsal ectoderm (reviewed in Ref. 14). Within the mesoderm, the dorsoventral gradient of Bmp specifies distinct mesodermal fates based on the level of Bmp signaling present, such that lower levels of Bmp signaling specify axial mesoderm more dorsally, which gives rise to the notochord, and paraxial mesoderm

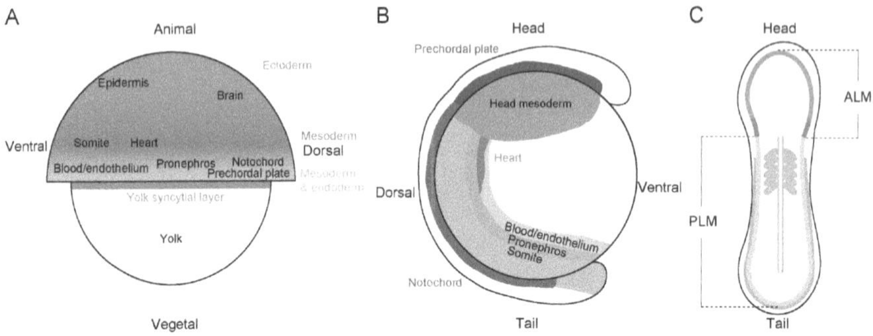


**Figure 5.1** The zebrafish fate map. (A) Fate map during gastrulation at 50% epiboly stage. Animal pole up, vegetal pole down, future dorsal side right, future ventral side left. Mesodermal fates are aligned along the dorsoventral axis. (B) Mesodermal fate model during early somitogenesis. Dorsal left, ventral right. Anterior up, posterior down. (C) Model of the lateral plate meosderm during early somitogenesis, dorsal projection, anterior up, posterior down. ALM, anterior lateral mesoderm; PLM, posterior lateral mesoderm. *After Ref. 7.* (See the color plate.)

more ventrally, which gives rise to the somites (Fig. 5.1A and B). Higher Bmp levels ventrally specify lateral plate mesoderm, from which zebrafish blood and endothelium are derived.[12] Furthermore, the level of Bmp signaling within the lateral plate mesoderm itself is an important determinant of fate specification within this tissue such that at the lowest levels of Bmp signaling, anterior lateral plate mesoderm (ALM) derivatives are formed, and heart, pronephros, and posterior lateral mesoderm (PLM) at the highest levels of Bmp signaling[15] (Fig. 5.1).

Transported via the cellular rearrangements and morphological movements of gastrulation, the lateral plate mesoderm eventually comes to reside as bilateral stripes found lateral and adjacent to the medial paraxial mesoderm, from which the somites develop (Fig. 5.1C). Bipotential cells able to form blood and endothelium, so-called hemangioblasts,[16] are present within the PLM and coexpress hematopoietic and endothelial genes until the 10-somite stage (Fig. 5.2A), at which point blood and endothelial gene expression diverge as the cells begin their medial migration, ventral to the somites and notochord[17] (Fig. 5.2B–D). Within the PLM, prior to and during migration, a hierarchy of lineage-specific differentiation genes is expressed.[18] Studies in the chicken embryo revealed that FGF signaling is required for the induction of angioblasts,[19,20] while a requirement for endodermally derived Hedgehog (Hh) signaling was demonstrated in vascular tube formation.[21] It is currently unclear whether FGF signaling is responsible for angioblast induction in zebrafish; however, a role for endoderm in this process has been excluded.[22] The zebrafish *cloche* mutant, named after

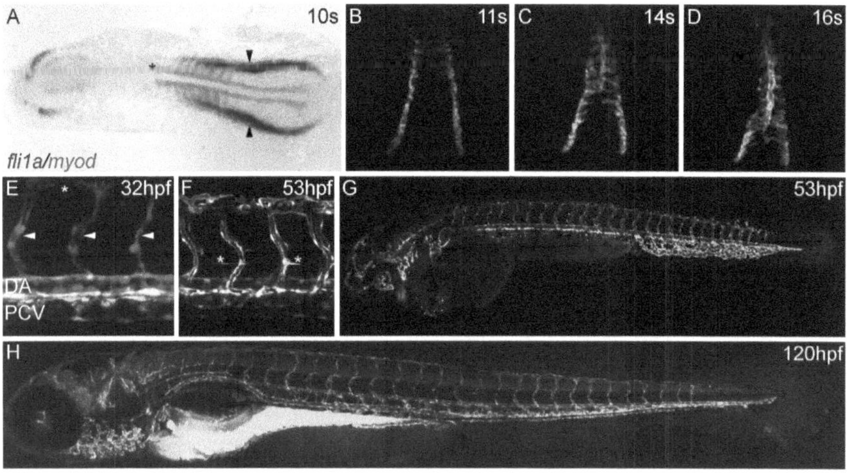


**Figure 5.2** Zebrafish vascular development. (A) *fli1a* expression within premigratory blood and vascular progenitors present in the posterior lateral mesoderm at 10-somite stage (black arrowheads). *myod* expression (asterisk) denotes the paraxial mesoderm and developing somites. (B–D) GFP expression within migratory angioblasts of Tg(*fli1a:EGFP*)*y1* embryos allow visualization of the early stages of vasculogenesis. (E) Tg(*fli1a:EGFP*)*y1*; *flt1enh:RFP* 32 hpf embryo depicting the DA and PCV, which are patent and the circulatory loop is well established. RFP-positive sprouting intersegmental arteries can be seen sprouting from the DA (arrowheads) where they migrate dorsally before sprouting laterally to form the DLAV (asterisk). (F) The parachordal vessel, the first major lymphatic vessel, begins to form after 2 days postfertilization (asterisks). (G, H) Tg(*fli1ep:mTQ2*)*sh321* at 2 and 5 days postfertilization, respectively, highlighting the rapid establishment of the complex vascular network. DA, dorsal aorta; DLAV, dorsolateral anastomotic vessel; PCV, posterior cardinal vein. (See the color plate.)

the French for bell, due to the bell-shaped heart in *cloche* embryos, fails to specify the hemangioblast progenitors of blood and endothelial lineages, leading to the loss of blood and blood vessels.[23] Since the *cloche* mutant displays such an early cardiovascular defect, its study has helped to identify many downstream proteins involved in both hematopoiesis and blood vessel formation. An acyl transferase, *lycat*, has been implicated as a candidate gene for *cloche*.[24] Overexpression of the transcription factor *etsrp* in zebrafish embryos rescues endothelial expression in *cloche* mutants, indicating that *etsrp* is a key transcriptional regulator sitting atop the endothelial hierarchy, which functions downstream of *cloche* to induce angioblast formation in zebrafish.[25]

## 3. ZEBRAFISH VASCULOGENESIS

The medial migration of angioblasts occurs in two distinct waves in an anterior–posterior direction (Fig. 5.2B–D). Lineage tracing studies have revealed that arterial and venous progenitors originate at distinct locations within the PLM.[26] The medial angioblasts, which are the first wave of angioblasts to migrate toward the midline, express the vascular endothelial growth factor (VEGF) receptor, *vegfr4/kdrl*, and the ETS-domain transcription factors, *fli1a* and *etsrp/etv2*.[26–28]

The first cells reach the midline at approximately 14–15-somite stage (Fig. 5.2C) and largely contribute to the embryonic artery, the dorsal aorta (DA), closely followed by a second wave containing primitive erythroid progenitors and angioblasts destined to form the posterior cardinal vein (PCV) (Fig. 5.2D).[27] Primitive erythroid progenitors, which express *gata1*, lie between the medial angioblasts and the lateral angioblasts, which largely contribute to the PCV.[26] The developing DA is comprised of several genetically distinct angioblast populations, deriving from different anterioposterior locations, and is differentially patterned by discrete chemokines.[29] By 18 s, the primitive erythroid progenitors and venous angioblasts present in the second migratory wave have reached the midline to form the intermediate cell mass (ICM), generating a midline endothelial cord. In contrast to the medially migrating angioblasts of the PLM, those angioblasts present within the ALM remain laterally positioned and form bilateral angioblast clusters via vasculogenesis. The cranial vessels of the zebrafish are subsequently generated from these clusters via angiogenic remodeling.[30]

The DA becomes patent between 22 and 23 hours postfertilization (hpf), prior to the onset of circulation, whereas the progenitors of the PCV and primitive erythrocytes remain intermingled for several hours. The PCV is not patent until the onset of circulation at approximately 26 hpf, when primitive erythrocytes begin to leave the ICM and enter circulation as the circulatory loop is established. Intersegmental vessels (ISVs), also referred to as intersegmental arteries (ISAs), begin sprouting from the DA by angiogenic remodeling of the primary vessel from approximately 23 hpf and are patent by 28 hpf (Fig. 5.2E, arrowheads).[31,32] Thus, the primary circulatory loop is established rapidly in the zebrafish embryo.

### 3.1. Regulation of initial angioblast migration

Multiple factors have been identified which influence the initial stages of vascular development. Zebrafish embryos in which the Hh pathway is activated exhibit impaired medial angioblast migration; however, Hh itself is unlikely to act as a midline attractant for migrating angioblasts, since *smo* mutant embryos, which lack all Hh signaling, exhibit relatively normal medial angioblast migration.[33] The role of Hh in this process is likely to be indirect, as the Hh receptors *ptch1* or *ptch2* are not detectably expressed within the PLM during somitogenesis,[34,35] which raises a question surrounding the identity of mediators of the Hh signal during angioblast migration. Repulsive signals, such as semaphorins, have been implicated in zebrafish angioblast migration; one example, *sem3a1/sema3aa*, which is expressed within the somites bordering the angioblast migratory path, is repressed by Hh signaling and is required for successful medial migration of DA angioblasts.[36,37]

While VEGF has been implicated in DA angioblast migration in organisms such as *Xenopus laevis* and is known to act downstream of Hh in the vascular development in zebrafish and mammals,[28,38,39] it is not required for medial angioblast migration in zebrafish, as evidenced by targeted knockdown of *vegfa*, its downstream effector *plcg1*, or by VEGF receptor inhibition.[22,33,40] Furthermore, the presence of fibronectin deposition on the ventral side of migrating angioblasts suggests that this may provide a directional migratory cue.[22] Additionally, loss of *etsrp* results in impaired angioblast migration, which raises the possibility that this gene may be downregulated by elevated levels of Hh signaling.[25]

## 3.2. Regulation of endothelial cord formation and arterial–venous specification

Following their arrival at the midline, angioblasts coalesce to form a single vascular cord by 19.5 hpf (21-somite stage) and remodeling of this vascular cord leads to vascular lumenization by approximately 22 hpf.[41] Live imaging studies of zebrafish ISVs have revealed that the lumen of ECs assembles by the fusion of intercellular vacuoles.[42] Anastomosis and vessel fusion in ISVs occur via cell membrane invagination and cord hollowing, which raises the possibility that these processes may similarly be involved in lumenization of the primary trunk vasculature.[43] Immediately following the formation of a vascular cord at the midline, the first differentially expressed endothelial markers are observed as ECs undergo arterial–venous specification.[22,44] A genetic basis for arterial and venous differentiation was first established with the identification of Ephrin-B2 ligand and EphB4 receptor tyrosine kinase as markers of arteries and veins, respectively.[45] The interaction of ligand and receptor requires cell–cell contact and defines and maintains the complex boundaries between arterial and venous territories; however, neither are required for fate specification of ECs.[45,46] Between 21 and 23 hpf, a subpopulation of angioblasts at the midline within the developing

DA migrate ventrally and contribute to the PCV, thereby demonstrating a common vessel primordium for the DA and PCV.[41] While VEGF signaling is not required for the earlier medial migration of angioblasts, it is critical for this later ventral migration stage.[33,41] Within the common vessel primordium, venous-fated angioblasts expressing *ephb4* migrate ventrally, away from *ephrinb2a*-expressing angioblasts and toward the territory of the PCV, via a process limited by VEGF and Notch signaling.[41] The transcription factor *tbx20*, regulated by VEGF, plays an important role in the segregation of the endothelial cord into arterial and venous domains, since *tbx20* knockdown in zebrafish embryos results in a single lumenized vessel at the embryonic midline.[47]

The Notch signaling pathway is a key determinant in the establishment of arterial identity.[48] In mice, the Notch1, Notch3, and Notch4 receptors, in addition to the Delta-like 4 (Dll4), Jagged1, and Jagged2 ligands, exhibit arterially restricted expression.[49] Furthermore, targeted inactivation of Notch1, Notch4, Dll4, and their targets Hey1 and Hey2 results in defective arterial specification (reviewed in Ref. 50). Cardiovascular diseases in humans such as Alagille syndrome and CADASIL syndrome (cerebral autosomal dominant arteriopathy with subcortical infarcts and leukoencephalopathy) have been shown to result from mutations in the Jagged1 ligand[51,52] and lesions within the Notch3 receptor, respectively.[53] Consistent with this, zebrafish mindbomb (*mib*) mutants, which are defective for Notch signaling, exhibit arterial–venous shunts, defective PCV formation, and reduced arterial gene expression, while gridlock (*grl*) mutants, which contain a lesion in *hey2*, exhibit reduced arterial gene expression and concomitant DA maturation defects.[48,54,55] Importantly, zebrafish *hey2/grl* is required upstream of Notch signaling for expression of arterial genes such as *ephrinb2a*[54] (Fig. 5.3).

While Notch signaling guides arterial fate, studies in zebrafish revealed that Hh and VEGF lie upstream of Notch in the process of arterial specification[28] (Fig. 5.3). This signaling hierarchy has also been confirmed to exist in mammals.[38] Zebrafish *sonic-you* (*syu*) mutant embryos, lacking midline Shha activity, display a strong reduction in arterial gene expression. The absence of paracrine Hh signaling from the midline leads to a failure to induce *vegfa* within the adjacent somites, which in turn results in a failure to induce Notch non-cell autonomously within angioblasts, thus resulting in a vessel primordium expressing venous and non-lineage restricted endothelial genes, but lacking arterial gene expression.[28] Vegfa binds to Vegfr2/Kdr or Vegfr4/Kdrl, or a receptor tyrosine kinase, thereby activating downstream Phospholipase C gamma 1 (Plcg1), which in turn activates

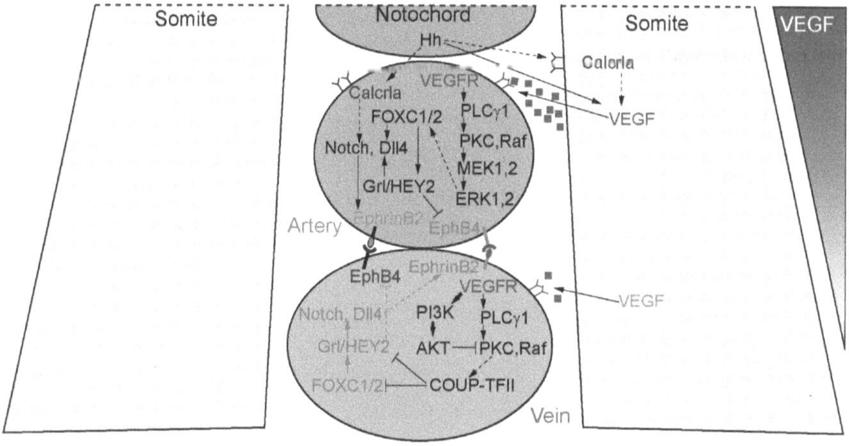


**Figure 5.3** Signaling hierarchies involved in arterial–venous differentiation. A model of the cellular signaling hierarchies and transcriptional programs is known to regulate artery and vein specification. For simplicity, the model depicts a schematic transverse section through a zebrafish embryo, with signaling pathways acting within arterial and venous ECs at the embryonic midline; however in reality, these signals also act earlier in migratory ECs. Signals highlighted in gray are opposed in the cell type in which they are shown. *After Ref. 44.* (See the color plate.)

ERK in the signal transduction cascade[40,56] (Fig. 5.3). In keeping with these observations, zebrafish *vegfr4* and *plcg1* mutants are defective in arterial, but not venous specification.[57,58]

Hedgehog signaling is known to induce expression of the Adrenomedullin receptor, Calcrla (Calcitonin receptor-like receptor alpha), within the somites and endothelial cord in zebrafish embryos.[59] In turn, Calcrla signaling induces *vegfa* expression downstream of Hh and upstream of Notch.[59] Hedgehog signaling is also able to induce somitic *vegfa* expression in a Calcrla-independent manner, and importantly, Hh can signal via Calcrla to induce arterial differentiation in a cell-autonomous VEGF-independent manner within angioblasts[33] (Fig. 5.3). In addition to *vegfr4*, the first migrating angioblasts to reach the midline also express activated phosphorylated ERK, and during later vasculogenesis this becomes restricted to arterial ECs.[44] Inhibition of an upstream activator of ERK, mitogen-activated or extracellular signal-related protein kinase kinase (MEK), leads to a specific loss of arterial differentiation, while inhibition of a parallel pathway, also downstream of VEGF, phosphatidylinositol-3-kinase (PI3K), leads to a specific loss of venous differentiation. Furthermore, these two pathways were shown to promote opposite EC fates, by the PI3K-mediated induction of

Akt, which is required to inhibit Raf, an upstream activator of ERK[44] (Fig. 5.3). Thus, a complex hierarchy of interacting signaling pathways functions in concert to establish arterial fate in ECs (Fig. 5.3).

At the transcriptional level, *foxc1a* and *foxc1b*, which are coexpressed in the nascent vasculature and somites, have been shown to act in concert to regulate arterial–venous differentiation.[60,61] Recent studies suggest that *foxc1a* and *foxc1b* may act in cooperation with *etsrp* at earlier stages of angioblast specification[60,62]; however, their mode of action during arteriovenous specification remains unclear. The mammalian orthologs of *foxc1a* and *foxc1b* are FoxC1 and FoxC2, respectively, and these genes are required for arterial specification in mammals via direct activation of components of the Notch signaling pathway (Fig. 5.3). However, *in vitro* studies suggest that this may occur via a mechanism modulated by VEGF signaling.[63,64]

The transcription factors *sox7* and *sox18* are known to act redundantly in arteriovenous differentiation, and double knockdown of *sox7* and *sox18* results in downregulation of arterial markers and ectopic expression of venous markers within the DA, leading to arteriovenous malformations.[65–67] Sox18 is also mutated in human hypotrichosis–lymphedema–telangiectasia syndrome, which results in reduced vascular integrity and hemorrhage.[68]

It has previously been suggested that whereas arterial fate is established by hierarchical Hh, VEGF, and Notch signaling,[28] venous fate is established via a default pathway arising from absent or low levels of these signaling hierarchies.[69] This view is now generally regarded as an oversimplification, since the orphan nuclear receptor, COUP-TFII/NR2F2, is specifically expressed within venous endothelium in the mouse and induces venous differentiation by actively inhibiting Notch signaling within the venous territory[70] (Fig. 5.3). In zebrafish, *nr2f2* expression is restricted to the PCV as early as 24 hpf (R.N. Wilkinson, unpublished observations) and *nr2f2* knockdown results in impaired venous development; however, it remains unclear whether this occurs as a result of activation of Notch signaling.[71] Inhibition of PI3K signaling in zebrafish embryos results in a loss of venous but not arterial ECs,[41,44] thereby illustrating that the regulation of venous identity is genetically controlled and not derived by a default signaling pathway (Fig. 5.3).

## 4. ANGIOGENESIS IN ZEBRAFISH

Although vasculogenesis is thought to be restricted to early development, most postnatal blood vessel formation occurs via the remodeling of existing vessels in a process known as angiogenesis. In zebrafish, arterial intersomitic vessels (aISVs) sprout bilaterally from the dorsal wall of the DA, anterior to each somite boundary from approximately 22 hpf.[32] ECs within the sprouting vessels are migratory, proliferative, and filopodia-rich, in contrast to ECs remaining within the trunk, termed "phalanx" ECs. During angiogenesis, the EC leading the sprouting vessel, known as the "tip cell," migrates, following guidance cues sensed by filopodia, while the trailing "stalk" ECs maintain their connection to the phalanx ECs of the vasculature and proliferate, forming the lumen of the vessel.[72,73] Tip cell migration and stalk cell proliferation can occur independently; however, both processes are tightly regulated by VEGF signaling.[74]

VEGF-A is a pivotal proangiogenic factor, which currently represents the major target for antiangiogenic therapy (reviewed in Ref. 75). In zebrafish, there are two VEGF-A paralogs, *vegfaa* and *vegfab*, both of which are required for normal angiogenesis; however, *vegfaa* plays a greater role.[28,76,77] In addition, *vegfc* loss of function also results in defective angiogenesis.[78,79] The output of VEGF is modulated by Dll4–Notch signaling, which itself limits angiogenesis within ECs and represents a switch by which angiogenic ECs are made unresponsive to VEGF[72,80] (Fig. 5.4). In the absence of Notch signaling, the behavioral distinction between tip and stalk ECs is lost, and increased numbers of ECs migrate into the endothelial sprout and proliferate. ECs within the sprout migrate away from the DA and exhibit increased filopodial activity.[72] This mechanism therefore controls the number of tip cells required for effective vascular network formation.

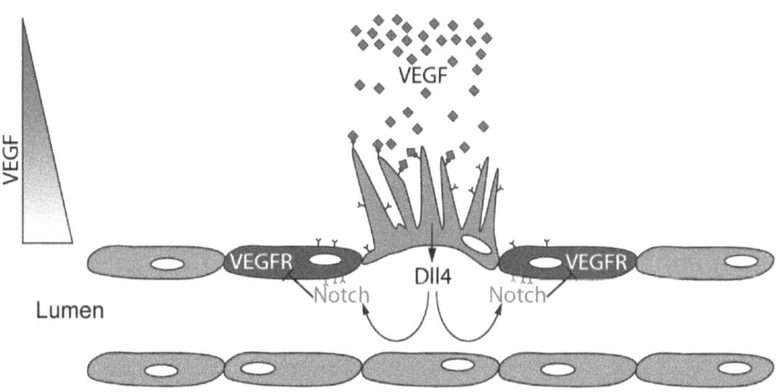


**Figure 5.4** Endothelial tip cell formation. VEGF is a pivotal proangiogenic signaling pathway whose output is modulated by Dll4–Notch signaling. A gradient of VEGF ligand is transduced by VEGF receptors present on ECs, which induces the expression of Dll4 within the tip cell. The induction of Dll4 acts as a molecular switch that via Notch signaling inhibits tip cell behavior in neighboring stalk cells, thereby limiting the ability of these ECs to respond to proangiogenic VEGF signals. (See the color plate.)

In zebrafish, filopodia are not required for EC tip cell guidance during angiogenesis, indicating these cells rely on alternative mechanisms to sense angiogenic signals in their environment.[81]

ECs dynamically compete with each other for the tip cell position via a process involving relative levels of the VEGF receptors Vegfr1/Flt1 (a VEGF decoy receptor) and Vegfr2/FLK1 (Vegfr4/Kdrl in zebrafish), such that ECs expressing higher levels of *Vegfr2* than *Vegfr1* have an increased chance of becoming a tip cell.[82] Elevated levels of *Vegfr2* lead to increased Dll4 expression, upstream of Notch signaling in ECs. Subsequently, Dll4-mediated activation of Notch signaling in adjacent stalk cells, in turn, inhibits *Vegfr2* and increases *Vegfr1* levels, thereby limiting the ability of neighboring ECs to respond to proangiogenic VEGF. Conversely, elevated levels of the decoy receptor *Vegfr1* reduce expression of Dll4 in ECs. The subsequent reduction of Notch signaling in neighboring ECs facilitates their integration at the leading edge of angiogenic sprouts, whereas increased Notch signaling restricts these cellular movements.[82] In keeping with this, downregulation of *vegfr1/flt1* expression in zebrafish embryos results in increased tip cell formation in a Notch-dependent manner.[83]

Paracrine Semaphorin–Plexin signaling is also required for proper aISV formation in zebrafish embryos.[73] Type 3 semaphorins are expressed within the somite, while their receptor, PlexinD1 is expressed within the developing vasculature. As described earlier, *sema3a1/sema3aa* is known to play a role in the regulation of early angioblast migration during primary vascular cord formation,[36] whereas a later role during angiogenesis has been identified for its paralog *sema3a2/sema3ab*.[73] Zebrafish embryos deficient in Sema–Plexin signaling exhibit ectopic vascular sprout formation, which is no longer restricted to segmental boundaries. This results in ectopic sprouts which display aberrant connections with their neighboring sprouts.[73] Sema–Plexin signaling exerts these antiangiogenic effects by antagonizing VEGF signaling via the upregulation of a VEGF decoy receptor, *sflt1*, thus limiting the VEGF signal received by ECs.[84]

Angiogenesis within veins is subject to different regulation than that occurring within arteries.[85,86] Venous ISVs sprout later than aISVs, from approximately 36 hpf onward. In zebrafish, the loss of *vegfr3/flt4* results in a failure of venous sprouting and lymphangiogenesis, while arterial angiogenesis remains unaffected.[85] Furthermore, the excessive arterial angiogenesis resulting from the loss of *dll4*[80] is rescued by the loss of *vegfr3* and increased by *vegfc* overexpression.[85] By revealing that a function of arterial Dll4 is to specifically limit the responsiveness of the DA to Vegfc–Vegfr3, these studies expand further the nature of Dll4 as a molecular switch regulating the responsiveness of ECs to VEGF signaling.[80,85] An additional level of regulation governing venous angiogenesis is provided by Bmp signaling and mediated by Disabled homologue 2 (Dab2), which specifically regulates venous but not arterial angiogenesis.[86,87] This additional signaling input indicates a potential mechanism by which overlapping arterial and venous angiogenic networks are permitted to develop independently.[86] Following sprouting angiogenesis, the vascular network undergoes extensive pruning through selective vessel regression to arrive at a mature vascular tree via a process coordinated by the Notch-regulated ankyrin repeat protein (Nrarp). Nrarp is induced by Dll4–Notch signaling in stalk ECs at newly formed branch points, where it regulates vessel density by integrating Notch and Wnt signaling to balance stalk cell proliferation.[88]

While the initial stages of aISV sprouting occur independently of blood flow, the onset of circulation provides a further level of regulation governing the development of the vasculature, namely, the physiological feedback mechanisms from oxygen transport and hemodynamic forces resulting from blood flow (reviewed in Ref. 89). The aISVs which sprout from the DA continue their dorsal migration until they reach a level which approximates that of the roof of the neural tube, at which point they bifurcate along the anteroposterior axis to anastomose with ipsilateral sprouts and form two independent bilateral dorsolateral anastomotic vessels lacking connections between them[32] (Fig. 5.2E, asterisk). This process has completed by 32 hpf, following which additional segmental sprouts form contralateral connections between the bilateral vessels to generate a plexus.[90] DLAV plexus formation requires both circulatory flow and VEGF signaling, which act in an additive manner to regulate its formation.[90] Angiogenic sprouting of a particular blood vessel, the accessory fifth aortic arch, has been shown to require the hemodynamic induction of a genetic pathway, where flow-mediated induction of the transcription factor *klf2a* induces an endothelial-specific microRNA, *mir-126*, which in turn activates proangiogenic VEGF signaling via repression of *spred1*.[91] Hemodynamic regulation of Cxcl12b–Cxcr4a chemokine signaling also regulates the formation of vascular networks within the developing central nervous system.[92,93]

## 5. LYMPHANGIOGENESIS IN ZEBRAFISH

Although blood vessels and lymphatic vessels are structurally related and function in concert, distinct molecular mechanisms are involved in their development. In contrast to blood vessels, which form a closed circulatory system, lymphatic vessels are blind ended and their primary function is to return fluid, macromolecules, and patrolling lymphocytes from interstitial spaces back into the blood circulation. As described earlier, the development of blood vessels proceeds via the mechanisms of vasculogenesis and angiogenesis, whereas lymphatic vessels form via a sprouting mechanism, termed lymphangiogenesis. Whereas the aISVs sprout from the DA at approximately 1 day per fertilization (dpf), the lymphatic network is generated slightly later from 2 dpf onward via angiogenic sprouting from the PCV, in keeping with the venous origin of the mammalian lymphatic vasculature.[32,94] The sprouting parachordal lymphangioblasts which originate from the PCV migrate dorsally and align along the horizontal myoseptum (Fig. 5.2F, asterisks), where they give rise to the thoracic duct, the intersegmental lymphatic vessels, and the dorsal longitudinal lymphatic vessel.[95] The migrating parachordal lymphangioblasts, which align along the horizontal myoseptum, are referred to as the parachordal vessel or parachordal chain. Once these cells have begun to migrate either dorsally or ventrally from the horizontal myoseptum, they become known as lymphatic endothelial cells (LECs). In a subset of the myotomal segments, the LECs connect to the primary aISV sprouts and remodel into venous ISVs, while some cells continue to migrate more dorsally and align along the horizontal myoseptum before recommencing migration either dorsally to give rise to the intersegmental lymphatic vessels or ventrally to contribute to the thoracic duct.[32,95]

In zebrafish, aISVs provide a guidance cue for migrating LECs; thus, abolition of aISV sprouting leads to a failure in the dorsal migration of LECs.[96] The nature of these guidance cues has been identified as the chemokine ligands *cxcl12a/sdf1a* and *cxcl12b/sdf1b* expressed along the LEC migratory path and the chemokine receptors *cxcr4a* and *cxcr4b*, which are expressed in migrating LECs.[97] Interestingly, it has been demonstrated that motoneurons present along the zebrafish trunk also guide the migration of LECS and formation of the parachordal chain and ultimately the thoracic duct.[98] Muscle pioneer cells lining the horizontal myoseptum express a guidance factor *netrin1a*, required for the formation of motor axons at the horizontal myoseptum. Loss of either *netrin1a* or its receptor *dcc* results in a loss of motor axons and failure to form the parachordal chain. Furthermore, laser ablation of motor axons along the horizontal myoseptum also prevents the formation of the parachordal chain and the lymphatic duct, indicating that motor axons themselves provide necessary guidance cues for the formation of the early lymphatic vasculature.[98] Whereas angiogenesis in blood vessels is orchestrated by Vegfa, lymphangiogenesis is regulated by Vegfc and its major receptor Vegfr3/Flt4.[99,100] As described earlier, loss of *vegfc*[79,100] but also loss of *vegfr3*[85] in zebrafish embryos results in a failure of lymphatic sprouting from the PCV. This is consistent with the finding that mutations in human VEGFR3/VEGFC have been demonstrated to cause primary lymphedema in multiple diseases including Milroy's disease.[101–104]

In mice, the sprouting parachordal lymphangioblasts which emerge from the PCV express the transcription factor *Prox1*.[105] Loss of *Prox1* results in a failure to form lymphatic vasculature, arising from defective LEC migration.[105] The expression of *Prox1* is induced by the transcription factor Sox18,[106] which also induces the expression of Nr2f2.[107] The interaction between Prox1 and Nr2f2 is critical to maintain Prox1 expression during the early stages of LEC migration. Zebrafish possess two paralogous *prox1* genes, *prox1a* and *prox1b*, of which *prox1b* is not required for lymphatic development.[108] As in mammals, zebrafish *prox1a* is initially expressed in migrating LECs and later throughout the lymphatic vasculature[109]; however, the requirement for *prox1a* during zebrafish lymphatic development remains controversial due to perceived indirect developmental defects following knockdown of *prox1a* expression.[100,108,109] In zebrafish, the secreted protein Ccbe1 is required for the sprouting, migration, and proliferation of lymphangioblasts; consequently, *ccbe1* mutants fail to properly form the lymphatic vasculature, resulting in severe lymphedema.[95] Mutations in human *Ccbe1* have recently been identified in Hennekam syndrome, which results in primary lymph-vessel dysplasia in humans,[110] further illustrating the conservation of mechanisms of lymphangiogenesis between zebrafish and humans.

## 6. VASCULAR MYOGENESIS

The maturation of blood vessels is accompanied by the recruitment of periendothelial cells, known as mural cells, which form contractile layers surrounding the vessels. Different periendothelial cell types are recruited depending upon the size and fate of the vessel; for example, pericytes are recruited to veins and capillaries, while vascular smooth muscle cell precursors migrate and surround developing arteries. The process of smooth muscle differentiation is termed vascular myogenesis. Vascular smooth muscle serves to change the caliber and thus the volume of blood vessels, thereby regulating local blood pressure. Smooth muscle cells are very plastic and able to act highly responsively to balance their cellular state between proliferation and differentiation in response to signaling inputs or to tissue injury.

Relatively little is known about vascular myogenesis in zebrafish, in comparison to mammals. Zebrafish mural cells expressing *smooth muscle alpha-actin* (*sm22alpha-b*), also known as *transgelin*, are first observed in a perivascular location from 72 hpf onward, surrounding the anterior DA and lateral DA.[153] Interestingly, *cloche* mutants, which lack all blood and endothelial progenitors,[23] retain Transgelin-positive cells in the location of the absent DA, while *hand2* mutants, which are defective in lateral mesoderm derived tissues, were devoid of such cells.[153] These observations indicate that zebrafish mural cells are derived from posterior lateral mesoderm and not from transdifferentiation of ECs; however, lineage-tracing studies remain to be conducted to establish the precise origin of these cells in zebrafish.

# 7. ZEBRAFISH AS A MODEL FOR VASCULAR DISEASE

## 7.1. HHT/arteriovenous malformations

Arteriovenous malformations are abnormal connections between arteries and veins, which manifest as twisted and often rupture-prone vessels. While many such malformations are present without symptoms, the size and location of the lesion determines the clinical severity, which can range from localized hypoxia to stroke in severe cases. The zebrafish is an emerging model for the study of arteriovenous malformations such as type 2 HHT, an autosomal dominant disorder with an incidence of 1 in 10,000. Type 2 HHT is caused by mutation of the TGF-β family type I receptor ACVRL1 or Alk1.[111] Other forms of HHT are caused by lesions within additional components of TGF-β family signal transduction, namely, a coreceptor, Endoglin, and a transcriptional mediator, Smad4.[112,113] The zebrafish mutant *violet beauregarde* was found to contain a lesion within the *alk1* gene.[114,115] As suggested by its name, *violet beauregarde* (*vbg*) mutant embryos develop severe edema, associated with an abnormal blood circulation, where the majority of blood flows through a limited number of dilated cranial vessels which contain increased numbers of ECs. No circulating blood is observed throughout the trunk vessels of *vbg* mutant embryos.[115] The presence of cerebral arteriovenous malformations in human type 2 HHT patients indicates the zebrafish may be a reliable model to study this disease. Arteriovenous malformations present in *vbg* mutants arise from the retention of normally transient arteriovenous connections in response to increased blood flow within dilated vessels.[116] Endothelial expression of *alk1* is induced prior to the onset of circulation (R.N. Wilkinson, unpublished observations); however, later increases within the cranial vasculature become flow dependent, suggesting that arteriovenous malformations within *vbg* mutants may represent an adaptive response to increased flow, by which the circulatory system remodels to drain the distended arterial system.[116]

In mammals, Alk1 was shown to signal via the Bmp branch of the TGF-β signaling pathway via Smad1/8 and this interaction is also conserved in zebrafish.[115,117] However, the TGF-β signaling components, TGFBR2 and ALK5, are not required for Alk1 activity within the endothelium, indicating that TGF-β subfamily proteins may not be the ligands for Alk1 pertinent to HHT pathogenesis.[118] Recent studies have implicated Bmp9 and Bmp10 as the ligands for Alk1.[119,120] Knockdown of the TGF-β-activated kinase, TAK1 also known as Map3k7, results in arteriovenous malformations of striking similarity to those observed in *alk1* mutants.[121] Furthermore, overexpression of *map3k7* can rescue arteriovenous malformations in an *alk1* loss-of-function background, thereby indicating that *map3k7* lies downstream of *alk1* signaling in vascular development.[121]

## 7.2. Klippel–Trenaunay syndrome

Klippel–Trenaunay syndrome is both a congenital and sporadic condition characterized by malformations in venous and lymphatic vessels, which results in hypertrophy of affected tissues.[122] Arterial vessels are unaffected in Klippel–Trenaunay patients. Multiple chromosomal abnormalities have been identified in Klippel–Trenaunay patients, but at present only one has been characterized, revealing a chromosomal translocation within the promoter region of AGGF1 (Angiogenic factor with G-patch and Forkhead-associated domain 1), which causes increased AGGF1 expression.[123] AGGF1 is secreted from ECs and is required for endothelial tube formation in matrigel assays.[123] In zebrafish, AGGF1 is expressed ubiquitously during early development and, when overexpressed, induces excessive ISA sprouting, abnormal PCV development, and excessive sprouting of subintestinal veins.[124] Conversely, knockdown of AGGF1 resulted in impaired angiogenic sprout formation in ISAs.[124] Arterial gene expression was unaffected following AGGF1 knockdown; however, venous expression of *dab2*, *flt4*, and *ephb4* was downregulated in the PCV, suggesting that venous differentiation is impaired in AGGF1 morphants.[124] The reduction of venous gene expression and defects in ISA sprouting could be partially rescued by overexpression of constitutively active AKT, indicating that AGGF1 may lie upstream of AKT during these processes.[124]

## 7.3. Cerebral cavernous malformations

CCMs represent vascular abnormalities consisting of clusters of dilated, thin-walled vessels, which are prone to hemorrhage. The disorder was named based on its occurrence within the blood vessels of the central nervous system although CCM can occur anywhere in the body and are distinct from arteriovenous malformations. The majority of CCMs are sporadic; however, familial forms occur, and taken together, these occur with remarkably high incidence of 1 in 250 individuals, although only 1 in 3 affected individuals are symptomatic.[125,126]

Currently, three genes have been identified which are linked to the onset of CCM; KRIT1 (*KREV1 interaction trapped 1*), which is also known as CCM1[127,128]; OSM (*Osmosensing scaffold for MEKK3*) also known as CCM2[129]; and PDCD10 (*Programmed cell-death 10*), also known as CCM3.[130] These proteins are known to interact with each other *in vitro* within a cytoskeletal and cell-junction-associated complex (reviewed in Ref. 131). In zebrafish, the *santa* mutant contains a lesion within the orthologous *ccm1* locus, while the *valentine* mutant contains a mutation within *ccm2*.[132,133] Loss of *ccm1* in zebrafish leads to a severe dilation of major blood vessels, followed by a thinning of cell walls, without affecting the fate of ECs or their number. The role of *ccm1* in the regulation of cellular morphogenesis was shown to occur cell autonomously within the endothelium and *ccm2* mutants were found to have a similar phenotype to that observed in *ccm1* mutants.[132,134] Furthermore, epistatic analysis has confirmed *in vivo* that both Ccm1 and Ccm2 function within the same complex.[132] The *heart of glass* zebrafish mutant contains a lesion within the receptor *heg1* and shows strikingly similar enlarged cardiac phenotypes to *santa* and *valentine*.[134] Heg1 was found to interact with Ccm2 to regulate endothelial cell–cell junctions and maintain vascular integrity, and this genetic interaction was also required for vessel lumenization.[134] Zebrafish also possess a related gene, *ccm2-like* (*ccm2l*), which interacts with *ccm1* and when knocked down enhances the phenotype of *ccm1*, *ccm2*, and *heg1* inhibition.[135] Interestingly, zebrafish *ccm1* and *ccm2* mutants exhibit normal arteriovenous specification, which conflicts with data in CCM1 knockout mice.[136] It has been proposed that loss of arterial markers in CCM1 knockout mice may result from a secondary failure of circulation in CCM1-deficient vessels and that Ccm1 primarily interacts with the cytoskeleton to regulate endothelial tubulogenesis downstream of differentiation events.[132]

Double knockdown of *ccm3a* and *ccm3b* in zebrafish phenocopies the vascular defects present in *ccm1* and *ccm2* mutants.[137,138] However, reduction in *ccm3a/b* gene function also induces gross dilation and mispatterning of cranial vessels, a phenotype not observed in *ccm1* and *ccm2* mutants. Importantly, this phenotype was independent of blood flow.[138] CCM3 has been shown to interact with a member of the germinal-center kinase III family, STK25.[139] Knockdown of *stk25b* in zebrafish phenocopies the enlarged cerebral vasculature observed in *ccm3a/b* morphants; furthermore, overexpression of *stk25b* mRNA rescues this defect.[138] Taken together, these data support a model whereby CCM1/CCM2 and CCM3/STK25 represent distinct arms of the CCM pathway, which regulate the development and integrity of the cranial vasculature.

## 7.4. von Hippel–Lindau disease

The zebrafish has more recently been employed as an important animal model of von Hippel–Lindau (VHL) disease, a rare autosomal dominant condition which predisposes individuals to benign and malignant, highly vascularized tumors.[140,141] VHL disease is caused by a mutation within the VHL tumor suppressor, and resultant tumors include clear cell renal carcinomas and retinal hemangioblastomas (reviewed in Ref. 142). VHL plays an important role in the cellular response to hypoxia via the downregulation of hypoxia-inducible factors (HIFs). Thus, inactivating mutations in the *Vhl* gene cause an increase in hypoxic response genes, one of which is VEGF-A.[143] Loss of Vhl function elevates the HIF pathway via stabilization of HIF, which in turn upregulates VEGF signaling, leading to aggressive angiogenesis and resulting in highly vascularized tumors. Indeed, zebrafish *vhl* mutants possess dramatically increased VEGF signaling, which manifests as global ectopic angiogenesis restricted to blood vessels. The lymphatic vasculature remains unaffected in *vhl* mutants since *vegfc* and *flt4* expression levels are normal, in contrast to elevated *vegfaa*, *vegfab*, *kdr*, and *kdrl*.[140] Interestingly, ectopic angiogenesis is not observed in zebrafish embryos until 58 hpf, where it presents as an increase in the number of cranial blood vessels. By 3.5 dpf, aberrant angiogenesis becomes evident in the trunk as an increase in ISV sprouting (Fig. 5.5). *vegf* expression was not assayed until late stages of development, so it remains unclear if VEGF signaling is elevated at earlier stages. Nevertheless, the potent proangiogenic role of VEGF, when

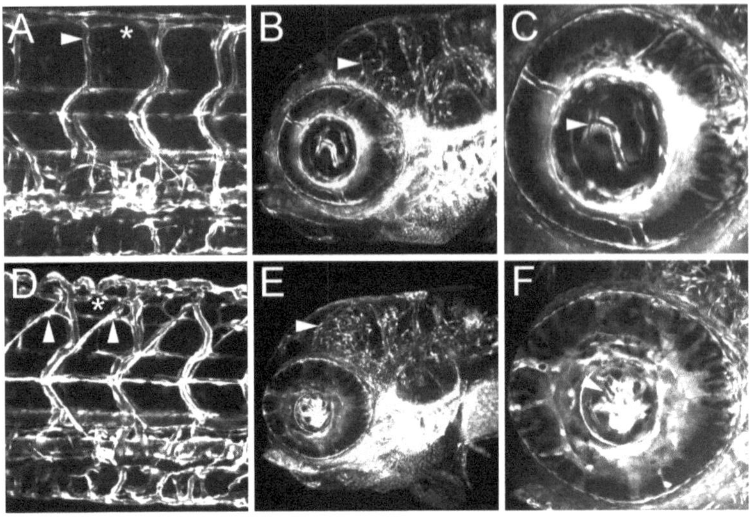


**Figure 5.5** Zebrafish von Hippel–Lindau (*vhl*) mutants exhibit hypoxic angiogenic defects. (A–C) Wild-type sibling at 5 dpf exhibiting normally patterned trunk vasculature (A), cranial vasculature (B), and retinal vasculature (C). (D–F) In contrast, *vhl* mutants exhibit abnormal ISV sprouting (A, D, arrowheads) and aberrant vessel looping within the DLAV (A, D, asterisks), increased cranial angiogenesis (B, E arrowheads), and increased retinal angiogenesis (C, F, arrowheads).

taken together with normal vascular development until 58 hpf, suggests that VEGF signaling is unresponsive to HIF signaling during early zebrafish development, as HIF reporters become activated around 24 hpf in this mutant. Interestingly, the ectopic angiogenesis present in *vhl* mutants could be rescued by stopping the blood circulation which led to a decrease in *dll4* expression and reduced Notch activity in ECs.[144] Importantly, these observations indicate that while Notch signaling is positively regulated by blood flow during early blood vessel development,[145] at later stages, blood flow acts to antagonize Notch via an as yet unidentified mechanism, thereby highlighting the context-dependent nature of hemodynamic regulation of angiogenesis.

## 7.5. Zebrafish as a model to study arterial occlusion and collateral vessel formation

Peripheral vascular disease, also known as peripheral artery occlusive disease, represents a narrowing or blockage of arteries. Arterial occlusion can occur via several mechanisms, including atherosclerotic plaque formation, thrombus formation, and also embolism, all of which obstruct arteries, restrict blood flow, and can lead to tissue ischemia. Ischemic tissue becomes hypoxic and ultimately results in localized tissue death. A common site for arterial occlusion to occur is within limbs, a situation that can often result in severe ischemia and necessitate amputation of the necrotic limb. The formation of collateral vessels to bypass the obstruction occurs via *de novo* remodeling of occluded arteries, in a process termed arteriogenesis.[146] Arteriogenesis represents an adaptive response to arterial obstruction and ultimately as a response to increased shear stress from blood flow.[147] Collateral vessels develop from preexisting arteriolar connections at the site of higher pressure, proximal to the occlusion, to connect at a site of lower pressure, distal to the occlusion. The use of the zebrafish to study atherosclerotic plaque formation has been described elsewhere[148]; therefore, this section focuses on the regulatory mechanisms employed by ECs to overcome obstructions within blood vessels.

By employing zebrafish as a system to study the process of vessel occlusion, novel insights into the regulatory mechanisms underlying arteriogenesis have been revealed. As described earlier, the zebrafish *gridlock* mutant possesses a mutation within the gene encoding *hey2*, a component of the Notch signaling pathway.[55] *gridlock* mutants display a permanent occlusion of the DA by approximately 48 hpf, such that there is no blood flow distal to the occlusion; however, the patterning of the remaining trunk vasculature remains normal.[149] Over subsequent days, distal blood flow is restored to the DA via mechanisms involving remodeling of existing vasculature, for example, via communications with the intestinal vasculature or by the reversal of flow within proximal ISAs.[149] Laser-induced arterial occlusion in zebrafish embryos achieved similar outcomes, with collateral aortic blood flow eventually restored distal to the site of occlusion.[149] While arteriogenesis often occurs at ischemic sites, *gridlock* embryos exhibited no upregulation of hypoxic markers. Thus, collateral aortic blood flow occurred in the absence of hypoxia and ischemia via a mechanism, which was shown to be dependent upon nitric oxide signaling.[149] Furthermore, the absence of blood flow has been shown to upregulate expression of the chemokine receptor *cxcr4a* in zebrafish ECs, and knockdown of *cxcr4a* in *gridlock* mutant embryos impairs collateral blood flow reestablishment.[150] Taken together, these data indicate that hemodynamic force rather than ischemia drives collateral vessel remodeling in zebrafish larvae.

Loss-of-function studies have revealed the involvement of parathyroid hormone receptor 1 (Pthr1) and its ligand parathyroid hormone-related peptide (Pthrp) in aortic occlusion in zebrafish embryos.[151]

Morpholino-mediated knockdown of *pthr1* and *pthrp* induce stereotypical localized occlusion of the mid-aorta in a Notch-dependent manner.[151] Interestingly, localized reductions in Notch signaling were observed in the hypochord, indicating that while *pthr1* is expressed in ECs, this effect may occur in a non-cell autonomous manner. Furthermore, Blomstrand chondroplasia, known to be caused by mutations in PTHR1,[152] often presents with accompanying aortic coarctation; thus, it is possible to speculate that defective PTHRP–PTHR1 signaling may account for this.

## 8. SUMMARY

The zebrafish is a powerful and highly versatile model in which the molecular mechanisms and cellular events that govern blood vessel formation are studied. The ability to directly visualize the formation of the vascular system *in vivo* in a non-invasive manner has led to numerous fundamental discoveries in the cell signaling and transcriptional hierarchies which regulate this process. The high level of evolutionary conservation between zebrafish and humans highlights the utility of zebrafish as a means of modeling human disease. With an increasing number of genes being implicated in human vascular disease and the recent advent of zebrafish-oriented targeted knockout technologies, the number of zebrafish models employed to study these diseases will increase substantially over the coming years. This can only serve to accelerate the study of vascular biology. The future certainly looks bright.

## ACKNOWLEDGMENTS

R. N. W. is supported by the JG Graves Medical Research Fellowship and the Royal Society (Research Grant RG120564). F. J. M. v. E. is supported by EC-FP7 HEALTH-F4-2010-242048.

## REFERENCES

1. Alitalo K. The lymphatic vasculature in disease. *Nat Med.* 2011;17(11):1371–1380.
2. Medvinsky A, Rybtsov S, Taoudi S. Embryonic origin of the adult hematopoietic system: advances and questions. *Development.* 2011;138(6):1017–1031.
3. Howe K, Clark MD, Torroja CF, et al. The zebrafish reference genome sequence and its relationship to the human genome. *Nature.* 2013;496(7446):498–503.
4. Siekmann AF, Covassin L, Lawson ND. Modulation of VEGF signalling output by the Notch pathway. *Bioessays.* 2008;30(4):303–313.
5. Herbert SP, Stainier DY. Molecular control of endothelial cell behaviour during blood vessel morphogenesis. *Nat Rev Mol Cell Biol.* 2011;12(9):551–564.

6. Mathieu J, Griffin K, Herbomel P, et al. Nodal and Fgf pathways interact through a positive regulatory loop and synergize to maintain mesodermal cell populations. *Development*. 2004;131(3):629–641.
7. Schier AF, Talbot WS. Molecular genetics of axis formation in zebrafish. *Annu Rev Genet*. 2005;39:561–613.
8. Walmsley M, Cleaver D, Patient R. Fibroblast growth factor controls the timing of Scl, Lmo2, and Runx1 expression during embryonic blood development. *Blood*. 2008;111(3):1157–1166.
9. Ben-Zvi D, Shilo BZ, Fainsod A, Barkai N. Scaling of the BMP activation gradient in Xenopus embryos. *Nature*. 2008;453(7199):1205–1211.
10. Dosch R, Gawantka V, Delius H, Blumenstock C, Niehrs C. Bmp-4 acts as a morphogen in dorsoventral mesoderm patterning in Xenopus. *Development*. 1997;124(12):2325–2334.
11. Marques G, Musacchio M, Shimell MJ, Wunnenberg-Stapleton K, Cho KW, O'Connor MB. Production of a DPP activity gradient in the early Drosophila embryo through the opposing actions of the SOG and TLD proteins. *Cell*. 1997;91(3):417–426.
12. Neave B, Holder N, Patient R. A graded response to BMP-4 spatially coordinates patterning of the mesoderm and ectoderm in the zebrafish. *Mech Dev*. 1997;62(2):183–195.
13. Tucker JA, Mintzer KA, Mullins MC. The BMP signaling gradient patterns dorsoventral tissues in a temporally progressive manner along the anteroposterior axis. *Dev Cell*. 2008;14(1):108–119.
14. Harland R, Gerhart J. Formation and function of Spemann's organizer. *Annu Rev Cell Dev Biol*. 1997;13:611–667.
15. Lieschke GJ, Oates AC, Paw BH, et al. Zebrafish SPI-1 (PU.1) marks a site of myeloid development independent of primitive erythropoiesis: implications for axial patterning. *Dev Biol*. 2002;246(2):274–295.
16. Sabin F. Studies on the origin of blood vessels and of red blood corpuscles as seen in the living blastoderm of chicks during the second day of incubation. *Carnegie Inst. Wash. Pub. # 272. Contrib Embryol.* 1920;9:214.
17. Brown LA, Rodaway AR, Schilling TF, et al. Insights into early vasculogenesis revealed by expression of the ETS-domain transcription factor Fli-1 in wild-type and mutant zebrafish embryos. *Mech Dev*. 2000;90(2):237–252.
18. Patterson LJ, Gering M, Patient R. Scl is required for dorsal aorta as well as blood formation in zebrafish embryos. *Blood*. 2005;105(9):3502–3511.
19. Cox CM, Poole TJ. Angioblast differentiation is influenced by the local environment: FGF-2 induces angioblasts and patterns vessel formation in the quail embryo. *Dev Dyn*. 2000;218(2):371–382.
20. Vokes SA, Krieg PA. Endoderm is required for vascular endothelial tube formation, but not for angioblast specification. *Development*. 2002;129(3):775–785.
21. Vokes SA, Yatskievych TA, Heimark RL, et al. Hedgehog signaling is essential for endothelial tube formation during vasculogenesis. *Development*. 2004;131(17):4371–4380.
22. Jin SW, Beis D, Mitchell T, Chen JN, Stainier DY. Cellular and molecular analyses of vascular tube and lumen formation in zebrafish. *Development*. 2005;132(23):5199–5209.
23. Stainier DY, Weinstein BM, Detrich 3rd HW, Zon LI, Fishman MC. Cloche, an early acting zebrafish gene, is required by both the endothelial and hematopoietic lineages. *Development*. 1995;121(10):3141–3150.
24. Xiong JW, Yu Q, Zhang J, Mably JD. An acyltransferase controls the generation of hematopoietic and endothelial lineages in zebrafish. *Circ Res*. 2008;102(9):1057–1064.
25. Sumanas S, Lin S. Ets1-related protein is a key regulator of vasculogenesis in zebrafish. *PLoS Biol*. 2006;4(1):e10.

26. Kohli V, Schumacher JA, Desai SP, Rehn K, Sumanas S. Arterial and venous progenitors of the major axial vessels originate at distinct locations. *Dev Cell.* 2013;25(2):196–206.
27. Gering M, Patient R. Hedgehog signaling is required for adult blood stem cell formation in zebrafish embryos. *Dev Cell.* 2005;8(3):389–400.
28. Lawson ND, Vogel AM, Weinstein BM. Sonic hedgehog and vascular endothelial growth factor act upstream of the Notch pathway during arterial endothelial differentiation. *Dev Cell.* 2002;3(1):127–136.
29. Siekmann AF, Standley C, Fogarty KE, Wolfe SA, Lawson ND. Chemokine signaling guides regional patterning of the first embryonic artery. *Genes Dev.* 2009;23(19):2272–2277.
30. Proulx K, Lu A, Sumanas S. Cranial vasculature in zebrafish forms by angioblast cluster-derived angiogenesis. *Dev Biol.* 2010;348(1):34–46.
31. Childs S, Chen JN, Garrity DM, Fishman MC. Patterning of angiogenesis in the zebrafish embryo. *Development.* 2002;129(4):973–982.
32. Isogai S, Horiguchi M, Weinstein BM. The vascular anatomy of the developing zebrafish: an atlas of embryonic and early larval development. *Dev Biol.* 2001;230(2):278–301.
33. Wilkinson RN, Koudijs MJ, Patient RK, Ingham PW, Schulte-Merker S, van Eeden FJ. Hedgehog signaling via a calcitonin receptor-like receptor can induce arterial differentiation independently of VEGF signaling in zebrafish. *Blood.* 2012;120(2):477–488.
34. Lewis KE, Concordet JP, Ingham PW. Characterisation of a second patched gene in the zebrafish Danio rerio and the differential response of patched genes to Hedgehog signalling. *Dev Biol.* 1999;208(1):14–29.
35. Williams C, Kim SH, Ni TT, et al. Hedgehog signaling induces arterial endothelial cell formation by repressing venous cell fate. *Dev Biol.* 2010;341(1):196–204.
36. Shoji W, Isogai S, Sato-Maeda M, Obinata M, Kuwada JY. Semaphorin3a1 regulates angioblast migration and vascular development in zebrafish embryos. *Development.* 2003;130(14):3227–3236.
37. Shoji W, Yee CS, Kuwada JY. Zebrafish semaphorin Z1a collapses specific growth cones and alters their pathway in vivo. *Development.* 1998;125(7):1275–1283.
38. Coultas L, Nieuwenhuis E, Anderson GA, et al. Hedgehog regulates distinct vascular patterning events through VEGF-dependent and -independent mechanisms. *Blood.* 2010;116(4):653–660.
39. Cleaver O, Krieg PA. VEGF mediates angioblast migration during development of the dorsal aorta in Xenopus. *Development.* 1998;125(19):3905–3914.
40. Lawson ND, Mugford JW, Diamond BA, Weinstein BM. Phospholipase C gamma-1 is required downstream of vascular endothelial growth factor during arterial development. *Genes Dev.* 2003;17(11):1346–1351.
41. Herbert SP, Huisken J, Kim TN, et al. Arterial-venous segregation by selective cell sprouting: an alternative mode of blood vessel formation. *Science.* 2009;326(5950):294–298.
42. Kamei M, Saunders WB, Bayless KJ, Dye L, Davis GE, Weinstein BM. Endothelial tubes assemble from intracellular vacuoles in vivo. *Nature.* 2006;442(7101):453–456.
43. Herwig L, Blum Y, Krudewig A, et al. Distinct cellular mechanisms of blood vessel fusion in the zebrafish embryo. *Curr Biol.* 2011;21(22):1942–1948.
44. Hong CC, Peterson QP, Hong JY, Peterson RT. Artery/vein specification is governed by opposing phosphatidylinositol-3 kinase and MAP kinase/ERK signaling. *Curr Biol.* 2006;16(13):1366–1372.
45. Wang HU, Chen ZF, Anderson DJ. Molecular distinction and angiogenic interaction between embryonic arteries and veins revealed by ephrin-B2 and its receptor Eph-B4. *Cell.* 1998;93(5):741–753.

46. Gerety SS, Wang HU, Chen ZF, Anderson DJ. Symmetrical mutant phenotypes of the receptor EphB4 and its specific transmembrane ligand ephrin-B2 in cardiovascular development. *Mol Cell*. 1999;4(3):403–414.
47. Szeto DP, Griffin KJ, Kimelman D. HrT is required for cardiovascular development in zebrafish. *Development*. 2002;129(21):5093–5101.
48. Lawson ND, Scheer N, Pham VN, et al. Notch signaling is required for arterial-venous differentiation during embryonic vascular development. *Development*. 2001;128(19):3675–3683.
49. Villa N, Walker L, Lindsell CE, Gasson J, Iruela-Arispe ML, Weinmaster G. Vascular expression of Notch pathway receptors and ligands is restricted to arterial vessels. *Mech Dev*. 2001;108(1–2):161–164.
50. Lin FJ, Tsai MJ, Tsai SY. Artery and vein formation: a tug of war between different forces. *EMBO Rep*. 2007;8(10):920–924.
51. Loomes KM, Underkoffler LA, Morabito J, et al. The expression of Jagged1 in the developing mammalian heart correlates with cardiovascular disease in Alagille syndrome. *Hum Mol Genet*. 1999;8(13):2443–2449.
52. Jones EA, Clement-Jones M, Wilson DI. JAGGED1 expression in human embryos: correlation with the Alagille syndrome phenotype. *J Med Genet*. 2000;37(9):658–662.
53. Joutel A, Corpechot C, Ducros A, et al. Notch3 mutations in CADASIL, a hereditary adult-onset condition causing stroke and dementia. *Nature*. 1996;383(6602): 707–710.
54. Rowlinson JM, Gering M. Hey2 acts upstream of Notch in hematopoietic stem cell specification in zebrafish embryos. *Blood*. 2010;116(12):2046–2056.
55. Zhong TP, Rosenberg M, Mohideen MA, Weinstein B, Fishman MC. Gridlock, an HLH gene required for assembly of the aorta in zebrafish. *Science*. 2000;287(5459):1820–1824.
56. Takahashi T, Shibuya M. The 230 kDa mature form of KDR/Flk-1 (VEGF receptor-2) activates the PLC-gamma pathway and partially induces mitotic signals in NIH3T3 fibroblasts. *Oncogene*. 1997;14(17):2079–2089.
57. Covassin LD, Siekmann AF, Kacergis MC, et al. A genetic screen for vascular mutants in zebrafish reveals dynamic roles for Vegf/Plcg1 signaling during artery development. *Dev Biol*. 2009;329(2):212–226.
58. Habeck H, Odenthal J, Walderich B, Maischein H, Schulte-Merker S. Analysis of a zebrafish VEGF receptor mutant reveals specific disruption of angiogenesis. *Curr Biol*. 2002;12(16):1405–1412.
59. Nicoli S, Tobia C, Gualandi L, De Sena G, Presta M. Calcitonin receptor-like receptor guides arterial differentiation in zebrafish. *Blood*. 2008;111(10):4965–4972.
60. De Val S, Chi NC, Meadows SM, et al. Combinatorial regulation of endothelial gene expression by ets and forkhead transcription factors. *Cell*. 2008;135(6):1053–1064.
61. Skarie JM, Link BA. FoxC1 is essential for vascular basement membrane integrity and hyaloid vessel morphogenesis. *Invest Ophthalmol Vis Sci*. 2009;50(11):5026–5034.
62. Veldman MB, Lin S. Etsrp/Etv2 is directly regulated by Foxc1a/b in the zebrafish angioblast. *Circ Res*. 2012;110(2):220–229.
63. Hayashi H, Kume T. Foxc transcription factors directly regulate Dll4 and Hey2 expression by interacting with the VEGF-Notch signaling pathways in endothelial cells. *PLoS One*. 2008;3(6):e2401.
64. Seo S, Fujita H, Nakano A, Kang M, Duarte A, Kume T. The forkhead transcription factors, Foxc1 and Foxc2, are required for arterial specification and lymphatic sprouting during vascular development. *Dev Biol*. 2006;294(2):458–470.
65. Cermenati S, Moleri S, Cimbro S, et al. Sox18 and Sox7 play redundant roles in vascular development. *Blood*. 2008;111(5):2657–2666.

66. Herpers R, van de Kamp E, Duckers HJ, Schulte-Merker S. Redundant roles for sox7 and sox18 in arteriovenous specification in zebrafish. *Circ Res.* 2008;102(1): 12–15.
67. Pendeville H, Winandy M, Manfroid I, et al. Zebrafish Sox7 and Sox18 function together to control arterial-venous identity. *Dev Biol.* 2008;317(2):405–416.
68. Irrthum A, Devriendt K, Chitayat D, et al. Mutations in the transcription factor gene SOX18 underlie recessive and dominant forms of hypotrichosis-lymphedema-telangiectasia. *Am J Hum Genet.* 2003;72(6):1470–1478.
69. Thurston G, Yancopoulos GD. Gridlock in the blood. *Nature.* 2001;414(6860):163–164.
70. You LR, Lin FJ, Lee CT, DeMayo FJ, Tsai MJ, Tsai SY. Suppression of Notch signalling by the COUP-TFII transcription factor regulates vein identity. *Nature.* 2005;435(7038):98–104.
71. Aranguren XL, Beerens M, Vandevelde W, Dewerchin M, Carmeliet P, Luttun A. Transcription factor COUP-TFII is indispensable for venous and lymphatic development in zebrafish and *Xenopus laevis*. *Biochem Biophys Res Commun.* 2011;410(1):121–126.
72. Siekmann AF, Lawson ND. Notch signalling limits angiogenic cell behaviour in developing zebrafish arteries. *Nature.* 2007;445(7129):781–784.
73. Torres-Vazquez J, Gitler AD, Fraser SD, et al. Semaphorin-plexin signaling guides patterning of the developing vasculature. *Dev Cell.* 2004;7(1):117–123.
74. Gerhardt H, Golding M, Fruttiger M, et al. VEGF guides angiogenic sprouting utilizing endothelial tip cell filopodia. *J Cell Biol.* 2003;161(6):1163–1177.
75. Schmidt T, Carmeliet P. Angiogenesis: a target in solid tumors, also in leukemia? *Hematology Am Soc Hematol Educ Program.* 2011;2011:1–8.
76. Bahary N, Goishi K, Stuckenholz C, et al. Duplicate VegfA genes and orthologues of the KDR receptor tyrosine kinase family mediate vascular development in the zebrafish. *Blood.* 2007;110(10):3627–3636.
77. Nasevicius A, Larson J, Ekker SC. Distinct requirements for zebrafish angiogenesis revealed by a VEGF-A morphant. *Yeast.* 2000;17(4):294–301.
78. Covassin LD, Villefranc JA, Kacergis MC, Weinstein BM, Lawson ND. Distinct genetic interactions between multiple Vegf receptors are required for development of different blood vessel types in zebrafish. *Proc Natl Acad Sci USA.* 2006;103(17):6554–6559.
79. Villefranc JA, Nicoli S, Bentley K, et al. A truncation allele in vascular endothelial growth factor c reveals distinct modes of signaling during lymphatic and vascular development. *Development.* 2013;140(7):1497–1506.
80. Leslie JD, Ariza-McNaughton L, Bermange AL, McAdow R, Johnson SL, Lewis J. Endothelial signalling by the Notch ligand Delta-like 4 restricts angiogenesis. *Development.* 2007;134(5):839–844.
81. Phng LK, Stanchi F, Gerhardt H. Filopodia are dispensable for endothelial tip cell guidance. *Development.* 2013;140(19):4031–4040.
82. Jakobsson L, Franco CA, Bentley K, et al. Endothelial cells dynamically compete for the tip cell position during angiogenic sprouting. *Nat Cell Biol.* 2010;12(10):943–953.
83. Krueger J, Liu D, Scholz K, et al. Flt1 acts as a negative regulator of tip cell formation and branching morphogenesis in the zebrafish embryo. *Development.* 2011;138(10):2111–2120.
84. Zygmunt T, Gay CM, Blondelle J, et al. Semaphorin-PlexinD1 signaling limits angiogenic potential via the VEGF decoy receptor sFlt1. *Dev Cell.* 2011;21(2):301–314.
85. Hogan BM, Herpers R, Witte M, et al. Vegfc/Flt4 signalling is suppressed by Dll4 in developing zebrafish intersegmental arteries. *Development.* 2009;136(23):4001–4009.

86. Wiley DM, Kim JD, Hao J, Hong CC, Bautch VL, Jin SW. Distinct signalling pathways regulate sprouting angiogenesis from the dorsal aorta and the axial vein. *Nat Cell Biol.* 2011;13(6):686–692.
87. Kim JD, Kang H, Larrivee B, et al. Context-dependent proangiogenic function of bone morphogenetic protein signaling is mediated by disabled homolog 2. *Dev Cell.* 2012;23(2):441–448.
88. Phng LK, Potente M, Leslie JD, et al. Nrarp coordinates endothelial Notch and Wnt signaling to control vessel density in angiogenesis. *Dev Cell.* 2009;16(1):70–82.
89. Liu D, Krueger J, Le Noble F. The role of blood flow and microRNAs in blood vessel development. *Int J Dev Biol.* 2011;55(4–5):419–429.
90. Zygmunt T, Trzaska S, Edelstein L, et al. 'In parallel' interconnectivity of the dorsal longitudinal anastomotic vessels requires both VEGF signaling and circulatory flow. *J Cell Sci.* 2011;125(pt 21):5159–5167.
91. Nicoli S, Standley C, Walker P, Hurlstone A, Fogarty KE, Lawson ND. MicroRNA-mediated integration of haemodynamics and Vegf signalling during angiogenesis. *Nature.* 2010;464(7292):1196–1200.
92. Bussmann J, Wolfe SA, Siekmann AF. Arterial-venous network formation during brain vascularization involves hemodynamic regulation of chemokine signaling. *Development.* 2011;138(9):1717–1726.
93. Fujita M, Cha YR, Pham VN, et al. Assembly and patterning of the vascular network of the vertebrate hindbrain. *Development.* 2011;138(9):1705–1715.
94. Srinivasan RS, Dillard ME, Lagutin OV, et al. Lineage tracing demonstrates the venous origin of the mammalian lymphatic vasculature. *Genes Dev.* 2007;21(19):2422–2432.
95. Hogan BM, Bos FL, Bussmann J, et al. Ccbe1 is required for embryonic lymphangiogenesis and venous sprouting. *Nat Genet.* 2009;41(4):396–398.
96. Bussmann J, Bos FL, Urasaki A, Kawakami K, Duckers HJ, Schulte-Merker S. Arteries provide essential guidance cues for lymphatic endothelial cells in the zebrafish trunk. *Development.* 2010;137(16):2653–2657.
97. Cha YR, Fujita M, Butler M, et al. Chemokine signaling directs trunk lymphatic network formation along the preexisting blood vasculature. *Dev Cell.* 2012;22(4):824–836.
98. Lim AH, Suli A, Yaniv K, Weinstein B, Li DY, Chien CB. Motoneurons are essential for vascular pathfinding. *Development.* 2011;138(17):3847–3857.
99. Karkkainen MJ, Haiko P, Sainio K, et al. Vascular endothelial growth factor C is required for sprouting of the first lymphatic vessels from embryonic veins. *Nat Immunol.* 2004;5(1):74–80.
100. Kuchler AM, Gjini E, Peterson-Maduro J, Cancilla B, Wolburg H, Schulte-Merker S. Development of the zebrafish lymphatic system requires VEGFC signaling. *Curr Biol.* 2006;16(12):1244–1248.
101. Irrthum A, Karkkainen MJ, Devriendt K, Alitalo K, Vikkula M. Congenital hereditary lymphedema caused by a mutation that inactivates VEGFR3 tyrosine kinase. *Am J Hum Genet.* 2000;67(2):295–301.
102. Karkkainen MJ, Ferrell RE, Lawrence EC, et al. Missense mutations interfere with VEGFR-3 signalling in primary lymphoedema. *Nat Genet.* 2000;25(2):153–159.
103. Brice G, Child AH, Evans A, et al. Milroy disease and the VEGFR-3 mutation phenotype. *J Med Genet.* 2005;42(2):98–102.
104. Gordon K, Schulte D, Brice G, et al. Mutation in vascular endothelial growth factor-C, a ligand for vascular endothelial growth factor receptor-3, is associated with autosomal dominant milroy-like primary lymphedema. *Circ Res.* 2013;112(6):956–960.
105. Wigle JT, Oliver G. Prox1 function is required for the development of the murine lymphatic system. *Cell.* 1999;98(6):769–778.
106. Francois M, Caprini A, Hosking B, et al. Sox18 induces development of the lymphatic vasculature in mice. *Nature.* 2008;456(7222):643–647.

107. Srinivasan RS, Geng X, Yang Y, et al. The nuclear hormone receptor Coup-TFII is required for the initiation and early maintenance of Prox1 expression in lymphatic endothelial cells. *Genes Dev.* 2010;24(7):696–707.
108. Tao S, Witte M, Bryson-Richardson RJ, Currie PD, Hogan BM, Schulte-Merker S. Zebrafish prox1b mutants develop a lymphatic vasculature, and prox1b does not specifically mark lymphatic endothelial cells. *PLoS One.* 2011;6(12):e28934.
109. Yaniv K, Isogai S, Castranova D, Dye L, Hitomi J, Weinstein BM. Live imaging of lymphatic development in the zebrafish. *Nat Med.* 2006;12(6):711–716.
110. Alders M, Hogan BM, Gjini E, et al. Mutations in CCBE1 cause generalized lymph vessel dysplasia in humans. *Nat Genet.* 2009;41(12):1272–1274.
111. Johnson DW, Berg JN, Baldwin MA, et al. Mutations in the activin receptor-like kinase 1 gene in hereditary haemorrhagic telangiectasia type 2. *Nat Genet.* 1996;13(2):189–195.
112. Gallione CJ, Richards JA, Letteboer TG, et al. SMAD4 mutations found in unselected HHT patients. *J Med Genet.* 2006;43(10):793–797.
113. McAllister KA, Grogg KM, Johnson DW, et al. Endoglin, a TGF-beta binding protein of endothelial cells, is the gene for hereditary haemorrhagic telangiectasia type 1. *Nat Genet.* 1994;8(4):345–351.
114. Driever W, Solnica-Krezel L, Schier AF, et al. A genetic screen for mutations affecting embryogenesis in zebrafish. *Development.* 1996;123:37–46.
115. Roman BL, Pham VN, Lawson ND, et al. Disruption of acvrl1 increases endothelial cell number in zebrafish cranial vessels. *Development.* 2002;129(12):3009–3019.
116. Corti P, Young S, Chen CY, et al. Interaction between alk1 and blood flow in the development of arteriovenous malformations. *Development.* 2011;138(8):1573–1582.
117. Chen YG, Massague J. Smad1 recognition and activation by the ALK1 group of transforming growth factor-beta family receptors. *J Biol Chem.* 1999;274(6):3672–3677.
118. Park SO, Lee YJ, Seki T, et al. ALK5- and TGFBR2-independent role of ALK1 in the pathogenesis of hereditary hemorrhagic telangiectasia type 2. *Blood.* 2008;111(2):633–642.
119. Wooderchak-Donahue WL, McDonald J, O'Fallon B, et al. BMP9 mutations cause a vascular-anomaly syndrome with phenotypic overlap with hereditary hemorrhagic telangiectasia. *Am J Hum Genet.* 2013;93(3):530–537.
120. Laux DW, Young S, Donovan JP, Mansfield CJ, Upton PD, Roman BL. Circulating Bmp10 acts through endothelial Alk1 to mediate flow-dependent arterial quiescence. *Development.* 2013;140(16):3403–3412.
121. Jadrich JL, O'Connor MB, Coucouvanis E. The TGF beta activated kinase TAK1 regulates vascular development in vivo. *Development.* 2006;133(8):1529–1541.
122. Wang QK. Update on the molecular genetics of vascular anomalies. *Lymphat Res Biol.* 2005;3(4):226–233.
123. Tian XL, Kadaba R, You SA, et al. Identification of an angiogenic factor that when mutated causes susceptibility to Klippel-Trenaunay syndrome. *Nature.* 2004;427(6975):640–645.
124. Chen D, Li L, Tu X, Yin Z, Wang Q. Functional characterization of Klippel-Trenaunay syndrome gene AGGF1 identifies a novel angiogenic signaling pathway for specification of vein differentiation and angiogenesis during embryogenesis. *Hum Mol Genet.* 2013;22(5):963–976.
125. Chan AC, Li DY, Berg MJ, Whitehead KJ. Recent insights into cerebral cavernous malformations: animal models of CCM and the human phenotype. *FEBS J.* 2010;277(5):1076–1083.
126. Riant F, Bergametti F, Ayrignac X, Boulday G, Tournier-Lasserve E. Recent insights into cerebral cavernous malformations: the molecular genetics of CCM. *FEBS J.* 2010;277(5):1070–1075.

127. Couteulx SL, Brezin AP, Fontaine B, Tournier-Lasserve E, Labauge P. A novel KRIT1/CCM1 truncating mutation in a patient with cerebral and retinal cavernous angiomas. *Arch Ophthalmol.* 2002;120(2):217–218.
128. Sahoo T, Johnson EW, Thomas JW, et al. Mutations in the gene encoding KRIT1, a Krev-1/rap1a binding protein, cause cerebral cavernous malformations (CCM1). *Hum Mol Genet.* 1999;8(12):2325–2333.
129. Liquori CL, Berg MJ, Siegel AM, et al. Mutations in a gene encoding a novel protein containing a phosphotyrosine-binding domain cause type 2 cerebral cavernous malformations. *Am J Hum Genet.* 2003;73(6):1459–1464.
130. Bergametti F, Denier C, Labauge P, et al. Mutations within the programmed cell death 10 gene cause cerebral cavernous malformations. *Am J Hum Genet.* 2005;76(1):42–51.
131. Storkebaum E, Quaegebeur A, Vikkula M, Carmeliet P. Cerebrovascular disorders: molecular insights and therapeutic opportunities. *Nat Neurosci.* 2011;14(11): 1390–1397.
132. Hogan BM, Bussmann J, Wolburg H, Schulte-Merker S. ccm1 cell autonomously regulates endothelial cellular morphogenesis and vascular tubulogenesis in zebrafish. *Hum Mol Genet.* 2008;17(16):2424–2432.
133. Mably JD, Chuang LP, Serluca FC, Mohideen MA, Chen JN, Fishman MC. Santa and valentine pattern concentric growth of cardiac myocardium in the zebrafish. *Development.* 2006;133(16):3139–3146.
134. Kleaveland B, Zheng X, Liu JJ, et al. Regulation of cardiovascular development and integrity by the heart of glass-cerebral cavernous malformation protein pathway. *Nat Med.* 2009;15(2):169–176.
135. Rosen JN, Sogah VM, Ye LY, Mably JD. ccm2-like is required for cardiovascular development as a novel component of the Heg-CCM pathway. *Dev Biol.* 2013;376(1):74–85.
136. Whitehead KJ, Plummer NW, Adams JA, Marchuk DA, Li DY. Ccm1 is required for arterial morphogenesis: implications for the etiology of human cavernous malformations. *Development.* 2004;131(6):1437–1448.
137. Voss K, Stahl S, Hogan BM, et al. Functional analyses of human and zebrafish 18-amino acid in-frame deletion pave the way for domain mapping of the cerebral cavernous malformation 3 protein. *Hum Mutat.* 2009;30(6):1003–1011.
138. Yoruk B, Gillers BS, Chi NC, Scott IC. Ccm3 functions in a manner distinct from Ccm1 and Ccm2 in a zebrafish model of CCM vascular disease. *Dev Biol.* 2012;362(2):121–131.
139. Voss K, Stahl S, Schleider E, et al. CCM3 interacts with CCM2 indicating common pathogenesis for cerebral cavernous malformations. *Neurogenetics.* 2007;8(4):249–256.
140. van Rooijen E, Voest EE, Logister I, et al. von Hippel-Lindau tumor suppressor mutants faithfully model pathological hypoxia-driven angiogenesis and vascular retinopathies in zebrafish. *Dis Model Mech.* 2011;3(5–6):343–353.
141. van Rooijen E, Voest EE, Logister I, et al. Zebrafish mutants in the von Hippel-Lindau tumor suppressor display a hypoxic response and recapitulate key aspects of Chuvash polycythemia. *Blood.* 2009;113(25):6449–6460.
142. Kaelin WG. The von Hippel-Lindau tumor suppressor protein: roles in cancer and oxygen sensing. *Cold Spring Harb Symp Quant Biol.* 2005;70:159–166.
143. Liu Y, Cox SR, Morita T, Kourembanas S. Hypoxia regulates vascular endothelial growth factor gene expression in endothelial cells. Identification of a 5′ enhancer. *Circ Res.* 1995;77(3):638–643.
144. Watson O, Novodvorsky P, Gray C, et al. Blood flow suppresses vascular Notch signalling via dll4 and is required for angiogenesis in response to hypoxic signalling. *Cardiovasc Res.* 2013;100(2):252–261.

145. Wang L, Zhang P, Wei Y, Gao Y, Patient R, Liu F. A blood flow-dependent klf2a-NO signaling cascade is required for stabilization of hematopoietic stem cell programming in zebrafish embryos. *Blood*. 2011;118(15):4102–4110.
146. Scholz D, Cai WJ, Schaper W. Arteriogenesis, a new concept of vascular adaptation in occlusive disease. *Angiogenesis*. 2001;4(4):247–257.
147. Pipp F, Boehm S, Cai WJ, et al. Elevated fluid shear stress enhances postocclusive collateral artery growth and gene expression in the pig hind limb. *Arterioscler Thromb Vasc Biol*. 2004;24(9):1664–1668.
148. Fang L, Liu C, Miller YI. Zebrafish models of dyslipidemia: relevance to atherosclerosis and angiogenesis. *Transl Res*. 2013;163(2):99–108.
149. Gray C, Packham IM, Wurmser F, et al. Ischemia is not required for arteriogenesis in zebrafish embryos. *Arterioscler Thromb Vasc Biol*. 2007;27(10):2135–2141.
150. Packham IM, Gray C, Heath PR, et al. Microarray profiling reveals CXCR4a is down-regulated by blood flow in vivo and mediates collateral formation in zebrafish embryos. *Physiol Genomics*. 2009;38(3):319–327.
151. Gray C, Bratt D, Lees J, et al. Loss of function of parathyroid hormone receptor 1 induces notch-dependent aortic defects during zebrafish vascular development. *Arterioscler Thromb Vasc Biol*. 2013;33(6):1257–1263.
152. Hoogendam J, Farih-Sips H, Wynaendts LC, Lowik CW, Wit JM, Karperien M. Novel mutations in the parathyroid hormone (PTH)/PTH-related peptide receptor type 1 causing Blomstrand osteochondrodysplasia types I and II. *J Clin Endocrinol Metab*. 2007;92(3):1088–1095.
153. Santoro MM, Pesce G, Stainier DY. Characterization of vascular mural cells during zebrafish development. *Mech Dev*. 2009;126(8–9):638–649. http://dx.doi.org/10.1016/j.mod.2009.06.1080. Epub 2009 Jun 17.

CHAPTER SIX

# Genetics of Response to Antiplatelet Therapy

**Mark R. Thomas, Robert F. Storey**
Department of Cardiovascular Science, University of Sheffield, Sheffield, United Kingdom

## Contents



*Progress in Molecular Biology and Translational Science*, Volume 124
ISSN 1877-1173
http://dx.doi.org/10.1016/B978-0-12-386930-2.00006-9

## Abstract

Dual antiplatelet therapy has a major role in the management of acute coronary syndromes (ACS) and following percutaneous coronary intervention (PCI). However, significant variation in pharmacodynamic response to antiplatelet therapy has been demonstrated, especially to clopidogrel. Single nucleotide polymorphisms, particularly those affecting the metabolism of antiplatelet therapy, account for some, but not all, of this variability in response.

Loss-of-function polymorphisms of CYP2C19, the gene encoding for the key enzyme in the metabolism of clopidogrel, are associated with reduced formation of the active metabolite of clopidogrel, a lower pharmacodynamic effect of the drug and a corresponding increase in adverse cardiovascular events. Conversely, gain-of-function polymorphisms of CYP2C19 are associated with an increased pharmacodynamic response to the drug and therefore an increase in bleeding. The clinical relevance of other polymorphisms that affect antiplatelet therapy has not been clearly established.

## 1. INTRODUCTION

Platelets play a pivotal role in the pathophysiology of coronary artery disease, which is the leading cause of mortality in developed countries. Although not solely responsible, they are a key mediator in the development of acute coronary syndromes (ACS) and their subsequent morbidity. Antiplatelet therapy has therefore become an essential addition to the armory of cardiologists for combating this disease, which is highlighted by their ranking as some of the most commonly prescribed drugs worldwide.

Dual antiplatelet therapy has developed an established role in the treatment of ACS and after percutaneous coronary intervention (PCI) to reduce the risk of adverse cardiovascular events. Current guidelines[1–4] recommend the use of a combination of aspirin and a platelet $P2Y_{12}$ ADP receptor antagonist (ticagrelor, prasugrel, or clopidogrel, depending on the clinical context) for the treatment of ACS or after PCI. Despite potent inhibition of platelet aggregation with dual antiplatelet therapy, the risk of further atherothrombosis is not entirely eliminated, for reasons that are complex and multifactorial. It has been demonstrated that there is great interindividual variability in pharmacodynamic response to clopidogrel in particular,[5] although variability in response to prasugrel has also been demonstrated.[6,7] A poor response to antiplatelet agents is associated with an increased risk of cardiovascular events.[8]

Genetic polymorphisms can, in part, explain some of the variability of pharmacokinetic and pharmacodynamic response to antiplatelet medications.

It is feasible that loss- or gain-of-function polymorphisms could manipulate key cellular mechanisms involved in metabolizing these medications and their mechanisms of action. This review summarizes the available evidence regarding the genetic component of response to antiplatelet agents.

## 2. ASPIRIN

The benefit of aspirin in the treatment of ACS and following the insertion of a coronary stent is well established.[9] Aspirin irreversibly inactivates the cyclooxygenase-1 (COX-1) enzyme. This prevents it from catalyzing the conversion of arachidonic acid to prostaglandins $G_2$ and $H_2$, thus inhibiting their conversion into the potent platelet agonist thromboxane $A_2$. This reduces the tendency of platelets to aggregate and form a thrombus.

### 2.1. Polymorphisms that affect response to aspirin

In total, more than 50 polymorphisms have been identified that could potentially affect response to aspirin. Of these, the $Pl^{A1/A2}$ polymorphism of GPIIIa and polymorphisms of *COX-1* have been the most extensively investigated. However, the findings of many of these studies have conflicted.

Polymorphisms of other genes have also been suggested to affect response to aspirin. These have mostly been genes that encode for platelet receptors such as GPIa, GPIbα, GPVI, $P2Y_1$, and $P2Y_{12}$. The clinical relevance of these polymorphisms has not been clearly established.

### 2.2. GPIIIa receptor polymorphisms

Polymorphisms have been identified in the GPIIIa subunit of the platelet GPIIb/IIIa receptor that mediates the cross-linking of platelets, resulting in platelet aggregation. One particular polymorphism, $Pl^{A2}$, has been associated with ACS.[10] However, analysis of the Physician's Health Study did not show carriage of the $Pl^{A2}$ allele to be a risk factor for the development of myocardial infarction (MI) in 14,916 prospectively observed healthy volunteers.[11]

*In vitro*, carriers of the $Pl^{A2}$ polymorphism show significantly greater ADP-induced platelet P-selectin expression, GPIIb/IIIa-bound fibrinogen, and activated GPIIb/IIIa determined by flow cytometry compared to non-carriers.[12] This suggests that the $Pl^{A2}$ polymorphism may be associated with hyperreactive platelets.

A systematic review by Goodman *et al.* published in 2008 analyzed 31 studies published before December 2007 to determine the effect of genetic polymorphisms on response to aspirin.[13] Combining the studies to analyze them was problematic due to significant heterogeneity between studies and the differing methodologies used to determine response to aspirin. The study concluded that the $Pl^{A1/A2}$ variant in the GPIIIa receptor gene was highly associated with poor response to aspirin in healthy subjects, although the strength of this association was not seen in patients with cardiovascular disease.

## 2.3. COX-1 polymorphisms

Multiple polymorphisms have been identified for *COX-1*, the gene that encodes for the enzyme targeted by aspirin. Although other polymorphisms have also been identified, the A-842G/C50T polymorphisms tend to have been the most studied. These polymorphisms are in complete linkage disequilibrium and are located within the promoter region and exon of *COX-1*, respectively. The A-842G/C50T haplotype may be associated with a greater response to aspirin compared to the common genotype, demonstrated by less formation of prostaglandin $H_2$ in healthy volunteers.[14] In contrast, two other studies suggested that the A-842G haplotype is associated with a lower response to aspirin than the common genotype, demonstrated by increased AA-induced platelet aggregation and thromboxane-$A_2$ formation in patients with coronary artery disease treated with aspirin. These polymorphisms have not, however, been demonstrated to affect the incidence of adverse cardiovascular events[15] or bleeding events[16] in patients treated with aspirin. Interpretation of these studies is difficult due to the small sample sizes. and the systematic review published by Goodman *et al.* did not find an association between *COX-1* genotype and response to aspirin.[13]

## 2.4. Summary of the effect of genotype on response to aspirin

There is limited evidence to suggest that carriage of $Pl^{A2}$ polymorphisms of the GPIIIa subunit may be associated with a reduced pharmacodynamic response to aspirin. However, none of the polymorphisms that potentially affect pharmacodynamic response to aspirin have been shown to be clinically relevant.

# 3. CLOPIDOGREL

Clopidogrel is a very commonly prescribed antiplatelet medication that was until recently the first line $P2Y_{12}$ antagonist for the treatment of ACS. Treatment with clopidogrel is currently recommended in addition to aspirin for patients undergoing PCI for stable coronary artery disease. There is substantial evidence for the benefit of clopidogrel compared to placebo for the treatment of non-ST-elevation ACS (NSTE-ACS)[17], ST-elevation ACS,[18] and following PCI.[19,20]

The first-generation thienopyridine, ticlopidine, has now been entirely replaced by clopidogrel, which is a second-generation thienopyridine. Clopidogrel is a prodrug that is converted *in vivo* into an active thiol-containing metabolite, which binds irreversibly to platelet ADP $P2Y_{12}$ receptors. Platelets are activated by agonists that are exposed or generated following atherosclerotic plaque rupture, such as collagen and thrombin. Upon activation, platelets release granule contents including potent platelet activators such as ADP, which results in an amplification of platelet activation by the initial stimulus. Blockade of the ADP receptors therefore interrupts this amplification pathway and reduces platelet reactivity and the tendency for thrombosis.

## 3.1. Metabolism of clopidogrel

Clopidogrel requires *in-vivo* conversion into its active metabolite in order to have an effect on platelet reactivity. After oral consumption, clopidogrel is absorbed by the intestinal mucosa, a process that is in part inhibited by the transportation of molecules across the intestinal epithelial cell membranes by P-glycoprotein (P-gp). The *ABCB1* (also known as *MDR1*) gene encodes for this efflux pump, which may affect the oral bioavailability of clopidogrel since increased expression of P-gp may impair absorption of clopidogrel into the circulation.[21] Approximately, 15% of the prodrug is then metabolized in two stages by cytochrome P450 (CYP) enzymes in the liver to become the clopidogrel active metabolite.[22] The remaining 85% of the prodrug is hydrolyzed by esterases, particularly hepatic carboxylesterase 1 (CES1), to become an inactive carboxylic acid derivative in a competing pathway. In the liver, clopidogrel is oxidized into 2-oxo-clopidogrel by CYP1A2, CYP2B6, and CYP2C19. This is then converted into clopidogrel's active metabolite by CYP2B6, CYP2C9, CYP2C19, and CYP3A4.[22] CYP2C19 is of particular importance as it has a major contribution to both of these metabolic steps. The key role of CYP2C19 in the metabolism of clopidogrel is demonstrated by the reduced pharmacodynamic effect of clopidogrel in the presence of inhibitors of CYP2C19.[23] There has also been a controversial suggestion that paraoxonase-1 (PON1) may be involved in the conversion of 2-oxo-clopidogrel to the final active metabolite,[24] although this has not been reproduced by subsequent studies and so has been convincingly refuted.[25–28]

## 3.2. Genetic polymorphisms and efficacy of clopidogrel

CYP2C19 has attracted considerable attention due to its role in both of the metabolic steps required for the production of clopidogrel. There is variability in the gene encoding this enzyme, and both loss-of-function and gain-of-function polymorphisms have been identified. There has been extensive investigation to determine whether or not CYP2C19 loss-of-function polymorphisms, in particular, explain why some patients have an incomplete response to clopidogrel. Although the majority of interest has focused on CYP2C19, studies have also investigated whether genetic variability in *ABCB1*, *PON1*, and *PEAR1* may also affect the efficacy of clopidogrel. It has been shown that genetic polymorphisms do not entirely explain the variability of response to clopidogrel. Even controlling for genetic polymorphisms that have been shown to affect response to antiplatelet therapy (CYP2C19, CYP3A5, *ABCB1*, and *PON1*), as well as known risk factors for high platelet reactivity, does not eliminate the variability in pharmacokinetic and pharmacodynamic response to clopidogrel.[27] It has been established that high pretreatment platelet reactivity is associated with high postclopidogrel platelet reactivity,[29] which may in part be related to significant variation in *P2RY12*, the gene encoding for the $P2Y_{12}$ receptor. Polymorphisms in *P2RY12* are associated with high on-clopidogrel platelet reactivity determined by light transmission aggregometry (LTA) and VerifyNow in patients undergoing PCI.[30]

## 3.3. CYP2C19 loss-of-function alleles and pharmacokinetics of clopidogrel

Multiple polymorphisms of the CYP2C19 gene have been identified. The CYP2C19*2, *3, *4, *5, *6, *7, *8, *9, and *10 polymorphisms confer a loss of function, while CYP2C19*17 is associated with a gain of function. The CYP2C19*2 polymorphism is the most common loss-of-function allele and therefore investigation has focused on this polymorphism. The prevalence of CYP2C19 loss-of-function polymorphisms is approximately 25–35% in the Caucasian population, but as high as 55–70% in Asian populations.[31] It is thought that a loss of function of CYP2C19 may lower the rate of conversion of clopidogrel into its thiol-containing active metabolite and result in more clopidogrel becoming inactivated by esterases. Healthy volunteer studies have consistently shown that carriage of CYP2C19 loss-of-function alleles is associated with up to 50% lower exposure to clopidogrel active metabolite, as determined by area-under-the-concentration-curve (AUC) and maximal plasma concentration.[26,32,33] It appears that this reduced exposure to clopidogrel active metabolite can be partially overcome by increasing the dose of clopidogrel.[34,35] Varenhorst *et al.* randomized 98 aspirin-treated patients with coronary artery disease to receive either clopidogrel or prasugrel.[36] Patients were then divided into two groups: carriers and noncarriers of CYP2C19 loss-of-function alleles. Carriers of CYP2C19 loss-of-function alleles who were treated with clopidogrel had significantly lower total plasma exposure to clopidogrel active metabolite, as determined by the AUC, than noncarriers ($p=0.0015$). Conversely, the presence of CYP2C19 loss-of-function alleles did not affect total exposure to prasugrel active metabolite ($p=0.636$). The variability in plasma concentrations of clopidogrel active metabolite has been shown to be the main determinant of the pharmacodynamic response of platelets to clopidogrel.[37,38]

## 3.4. CYP2C19 loss-of-function alleles and pharmacodynamic effect of clopidogrel

Pharmacodynamic response can be assessed by measurement of platelet reactivity, and various methodologies have been developed for this purpose. LTA, multiple electrode aggregometry (MEA), and VerifyNow essentially measure the degree of aggregation of platelets in response to a stimulus. The vasodilator-stimulated phosphoprotein (VASP) phosphorylation assay specifically assesses the activation of $P2Y_{12}$ receptor, since the phosphorylation of VASP is inhibited in response to activation of $P2Y_{12}$. Although definitions vary based on the test that is used, high on-treatment platelet reactivity determined by LTA, MEA, VASP, and VerifyNow is associated with adverse cardiovascular events.[8,39] After clopidogrel treatment, healthy volunteers who are carriers of the CYP2C19*2 loss-of-function allele have significantly higher on-treatment platelet reactivity determined by LTA,[32,40] VerifyNow,[26] and VASP phosphorylation.[40] Varenhorst *et al.*'s randomized study also investigated the effect of CYP2C19 loss-of-function alleles on the pharmacodynamic effect of clopidogrel and prasugrel.[36] Clopidogrel-treated patients who were carriers of CYP2C19 loss-of-function alleles had significantly higher on-treatment platelet reactivity determined by VerifyNow test and VASP phosphorylation. The effect of prasugrel on platelet reactivity was not affected by CYP2C19 genotype.

### 3.4.1 Summary of PK/PD effect of CYP2C19 loss-of-function polymorphisms

During clopidogrel therapy, carriers of CYP2C19 loss-of-function alleles have a lower exposure to clopidogrel active metabolite and a lower response to the drug, demonstrated by higher on-treatment platelet reactivity.

## 3.5. Effect of CYP2C19 loss-of-function alleles on clinical outcomes

### 3.5.1 Key observational studies of PCI and ACS patients

Two thousand four hundred and eighty-five clopidogrel-treated patients undergoing PCI have been genotyped for the CYP2C19*2 loss-of-function allele to determine its effect on stent thrombosis.[41] CYP2C19*2 carriers had a significantly higher rate of stent thrombosis than noncarriers (adjusted HR 3.81; 95% CI 1.45–10.02; $p=0.007$). In particular, homozygotes for CYP2C19*2 had the highest risk compared to noncarriers (HR 5.25; $p=0.002$).

Two thousand two hundred and eight clopidogrel-treated ACS patients have been genotyped for CYP2C19*2 and CYP2C19*3 loss-of-function alleles in the French registry of acute ST-elevation and non-ST-elevation Myocardial Infarction (FAST-MI).[42] In this study, carriage of two CYP2C19 loss-of-function alleles was associated with an increased risk of subsequent adverse cardiovascular events (adjusted HR 1.98; 95% CI 1.10–3.58), but carriage of one CYP2C19 loss-of-function allele was not (adjusted HR 0.69; 95% CI 0.51–0.93). The increased risk of cardiovascular events was particularly high in carriers of two CYP2C19 loss-of-function alleles who underwent PCI (HR 3.58; 95% CI 1.71–7.51).

Nine hundred and twenty-eight clopidogrel-treated ACS patients have been genotyped to determine the effect of the CYP2C19*2 allele on subsequent adverse cardiovascular events.[43] In contrast to the other studies, the risk of subsequent adverse cardiovascular events did not differ in patients with one or more CYP2C19*2 alleles compared to noncarriers.

### 3.5.2 Key subgroup analyses of ACS clinical trials (see Tables 6.1 and 6.2)

The genetic substudy of the PLATelet inhibition and patient Outcomes (PLATO) trial[44] is the largest single study to have assessed the influence of CYP2C19 genotype on clinical outcomes. Ten thousand two hundred and eighty-five ACS patients were genotyped for CYP2C19 loss-of-function alleles (**2, *3, *4, *5, *6, *7,* and **8*). In clopidogrel-treated patients ($n=5148$), the primary endpoint (CV death, MI, and stroke) occurred more frequently in the first 30 days in patients with loss-of-function alleles compared to those without (5.7% vs. 3.8%, $p=0.028$). The overall rate of the primary endpoint was numerically higher in clopidogrel-treated patients with a loss-of-function allele compared to those without, but this was not statistically significant (11.2% vs. 10.0%, $p=0.25$). Clopidogrel-treated patients with CYP2C19 loss-of-function alleles also had a numerically lower rate of non-CABG-related bleeding than noncarriers (3.2% vs. 3.6%), although this was also not statistically significant ($p$ for interaction$=0.31$).

The genetic substudy of the Trial to Assess Improvement in Therapeutic Outcomes by Optimizing InhibitioN with Prasugrel—Thrombolysis in

**Table 6.1** CYP2C19 loss-of-function alleles and adverse cardiovascular events in clopidogrel-treated patients in major ACS trials

| Source | Comparison | CYP2C19 LoF alleles genotyped | Primary outcome | Events | HR/RR | 95% CI |
|---|---|---|---|---|---|---|
| PLATO substudy[44,45] | Any LoF versus no LoF | 2–8 | CV death, MI, or stroke | 149/1388 (10.7%) versus 332/3516 (9.4%) | 1.14 | 0.95–1.37 |
| TRITON substudy[45,46] | Any LoF versus no LoF | 2–10 and 12–14 | CV death, MI, or stroke | 46/395 (11.6%) versus 83/1064 (7.8%) | 1.49 | 1.06–2.10 |
| CURE substudy[45,47] | Any LoF versus no LoF | 2 and 3 | CV death, MI, or stroke | 52/650 (8.0%) versus 178/1880 (9.4%) | 0.84 | 0.63–1.14 |

LoF, loss of function.

**Table 6.2** Interaction between CYP2C19 LoF alleles and effect of clopidogrel compared to placebo in major ACS trials

| Source | CYP2C19 | Clopidogrel | Placebo | HR | 95% CI |
|---|---|---|---|---|---|
| CURE substudy[45,47] | Any LoF | 52/650 (8.0%) | 78/673 (11.6%) | 0.69 | 0.49–0.96 |
| | No LoF | 178/1880 (9.5%) | 233/1813 (12.9%) | 0.74 | 0.61–0.89 |
| CLARITY substudy[45,48] | Any LoF | NR | NR | 0.40 | 0.15–1.10 |
| | No LoF | NR | NR | 0.55 | 0.25–1.05 |

LoF, loss of function; NR, not reported.

Myocardial Infarction (TRITON-TIMI 38) trial[46] genotyped 2932 ACS patients for loss-of-function alleles. Clopidogrel-treated patients who were carriers of a CYP2C19 loss-of-function allele had a higher incidence of the primary endpoint than noncarriers (HR 1.77, 95% CI 1.11–2.80, $p=0.0155$).

The genetic substudy of the Clopidogrel in Unstable Angina to Prevent Recurrent Events (CURE) trial[47] genotyped 5059 ACS patients for CYP2C19 loss-of-function alleles. Clopidogrel was associated with a reduced incidence of the primary endpoint (CV death, nonfatal MI, or stroke) compared to placebo in both carriers of CYP2C19 loss-of-function alleles (HR 0.69, 95% CI 0.59–0.87) and noncarriers (HR 0.72, 95% CI 0.59–0.87). In this study, there was no evidence that carriage of a CYP2C19 loss-of-function allele was associated with reduced efficacy of clopidogrel ($p=0.84$ for interaction). The effect of clopidogrel on major bleeding compared to placebo was also consistent in carriers and noncarriers of CYP2C19 loss-of-function alleles. However, in contrast to PLATO and TRITON, only 15.5% of patients underwent PCI in the CURE study and this may, in part, explain why CYP2C19 genotype appeared to have less influence on clinical outcome in this study.

### 3.5.3 Meta-analysis and systematic review

Five meta-analyses of studies published prior to early August 2010 were unanimous in concluding that CYP2C19 loss-of-function alleles were associated with an increased incidence of cardiovascular events in patients treated with clopidogrel.[49–53] These studies showed that carriers of CYP2C19 loss-of-function alleles had 30–96% higher risk of adverse cardiovascular events than noncarriers. Interestingly, the inclusion of data from the subsequent genetic substudies of CURE/ACTIVE A[47] and PLATO[44] in late August 2010 has resulted in three subsequent meta-analyses questioning the role of CYP2C19 loss-of-function alleles in clopidogrel-treated patients.[45,54,55]

Although previous meta-analyses had concluded that CYP2C19 loss-of-function alleles are associated with an increased incidence of adverse cardiovascular events, one of the largest and most recent meta-analyses questions this. This study is a meta-analysis of 32 studies (42,016 patients with cardiovascular disease) published before October 2011.[45] In studies where clopidogrel was not compared to placebo, CYP2C19 loss-of-function alleles were associated with an increased incidence of adverse cardiovascular events (RR 1.18, 95% CI 1.09–1.28). Carriers of CYP2C19 loss-of-function alleles also had a lower risk of bleeding events than noncarriers (RR 0.84; 95% CI 0.75–0.94).

There was evidence of small study bias in this meta-analysis and when attempts were made to correct for this, the increased risk associated with CYP2C19 loss-of-function alleles was attenuated. Using a theoretical model (trim-and-fill analysis) to account for small study bias reduced the RR to 1.10 (95% CI 1.02–1.19) for fixed effects and 1.13 (95% CI 0.96–1.33) for random-effects models. The RR associated with CYP2C19 loss-of-function alleles was diminished as the size of the studies increased. The overall RR in studies that reported 1–99 adverse cardiovascular events (16 studies) was determined to be 1.83 (95% CI 1.50–2.23). The RR in studies that reported 100–199 events (six studies) was 1.26 (95% CI 1.09–1.45) and the corresponding RR in studies that reported >200 events (four studies that accounted for over 50% of the total events in this meta-analysis) was 0.97 (95% CI 0.86–1.09).

This study also analyzed whether there was an interaction between carriage of CYP2C19 loss-of-function alleles and the effect of clopidogrel compared to placebo in four clinical trials (ACTIVE-A, CURE, CHARISMA, and CLARITY-TIMI 28). The effect of clopidogrel compared to placebo was lower in patients with CYP2C19 loss-of-function alleles (HR 0.87; 95% CI 0.70–1.09) than in patients without CYP2C19 loss-of-function alleles (HR 0.78; 95% CI 0.69–0.89) although this was not significant ($p=0.37$ for interaction).

This meta-analysis concluded that there was no significant association of CYP2C19 genotype with adverse cardiovascular events. This comprehensive meta-analysis was, however, limited by several issues. While the overall data from the meta-analysis showed that CYP2C19 loss-of-function alleles were associated with a significant increase in risk of cardiovascular events, this was disregarded because theoretical models, which accounted for small study bias, attenuated this increased risk. While small study bias will undoubtedly have influenced the association of CYP2C19 loss-of-function alleles with cardiovascular events, a theoretical model cannot perfectly correct for this and the corrected results should be interpreted with caution.

Particular importance was attributed to the four substudies of trials that compared the effect of clopidogrel with placebo, as these studies effectively control for any potential effects of CYP2C19 that are not related to clopidogrel metabolism. However, of these, two of the original studies showed marginal overall benefit of clopidogrel (CHARISMA and ACTIVE-A), and therefore, a decrease in the effect of clopidogrel related to CYP2C19 loss-of-function alleles would be expected to have negligible effects on the outcome of these patients. Another one of the substudies was incompletely reported and is currently unpublished and not peer reviewed (substudy of CLARITY).

Another limitation of the meta-analysis is the heterogeneity of the studies that it includes. The indication for clopidogrel was not consistent and included stable coronary artery disease, ACS, and thromboprophylaxis for atrial fibrillation. The clinical importance of the effect of CYP2C19 loss-of-function polymorphisms in patients with ACS may well have been overshadowed by the lesser modification of risk in lower-risk patients who would not be expected to benefit from clopidogrel to as great an extent.

### *3.5.4 Summary of the effect of CYP2C19 loss-of-function polymorphisms on clinical outcomes*

In keeping with the pharmacokinetic and pharmacodynamic data, meta-analyses have shown that CYP2C19 loss-of-function polymorphisms are associated with an increased incidence of adverse cardiovascular events in clopidogrel-treated patients. Small study bias is likely to have influenced the magnitude of the association however. CYP2C19 loss-of-function polymorphisms are likely to be most clinically relevant for high-risk patients treated with clopidogrel, such as ACS patients within the first 30 days of PCI, whereas there may be little or no impact of these polymorphisms in patients with stable coronary artery disease who have not recently undergone PCI.

## 3.6. CYP2C19 gain-of-function alleles

While the majority of CYP2C19 polymorphisms confer a loss of function, CYP2C19*17 is associated with a gain of function and an ultrarapid metabolizer phenotype. CYP2C19*17 alleles have a frequency of approximately 15–30% in Caucasian and African populations, but only approximately 0–5% in Chinese and Japanese populations.[56]

### *3.6.1 CYP2C19 gain-of-function alleles and the pharmacokinetic and pharmacodynamic response to clopidogrel*

In elective PCI patients treated with clopidogrel, there is a gene–dose effect relationship between carriage of CYP2C19*17 alleles and platelet reactivity determined by MEA.[57] In this study of 986 patients, carriers of at least one CYP2C19*17 allele had lower platelet reactivity and were less likely to be classified as low responders to clopidogrel than noncarriers. This was even more pronounced in homozygotes for CYP2C19*17 alleles. This study also showed that the loss of function associated with a CYP2C19*2 allele had a greater effect than the gain of function associated with a CYP2C19*17 allele in patients who were heterozygous for both loss- and gain-of-function alleles. A study of 598 clopidogrel-treated patients with NSTE-ACS demonstrated that carriage of CYP2C19*17 is significantly associated with lower platelet reactivity determined by VASP phosphorylation but not LTA.[58]

### *3.6.2 Summary of effect of CYP2C19*17 on pharmacodynamic effect of clopidogrel*

During clopidogrel therapy, carriers of the CYP2C19*17 gain-of-function allele have a higher pharmacodynamic response to the drug, demonstrated by lower platelet reactivity.

### *3.6.3 Key observational studies investigating clinical outcomes*

Clopidogrel-treated patients undergoing PCI who were carriers of the CYP2C19*17 allele had a greater risk of bleeding than noncarriers, in a study of 1524 patients.[59] After a clopidogrel loading dose, patients who were carriers of at least one CYP2C19*17 allele ($n=622$) had lower platelet reactivity determined by MEA than noncarriers ($p=0.039$). Homozygotes for CYP2C19*17 ($n=76$) had even lower platelet reactivity ($p=0.008$). Carriers of at least one CYP2C19*17 allele also had a higher risk of bleeding (TIMI major and minor bleeding) than noncarriers (OR 1.80, 95% CI 1.03–3.14) and the risk was highest among homozygotes for CYP2C19*17

(OR 3.27, 95% CI 1.33–8.10). In this study, CYP2C19*17 did not influence the occurrence of stent thrombosis.

A study of 928 clopidogrel-treated patients with ACS has demonstrated a lower risk of adverse cardiovascular events in carriers of CYP2C19*17, compared to noncarriers. Carriers of at least one CYP2C19*17 allele had a lower risk of ischemia-driven target vessel revascularization (OR 0.57, 95% CI 0.39–0.82, $p=0.002$) and a lower risk of adverse cardiovascular events compared to noncarriers (OR 0.72, 95% CI 0.53–0.98, $p=0.04$).[43]

In the FAST-MI registry,[42] carriage of CYP2C19*17 was not shown to influence the risk of adverse cardiovascular events in post-MI patients treated with clopidogrel.

### 3.6.4 Genetic subgroup analyses of large ACS clinical trials

In the PLATO study, 10,285 ACS patients were genotyped for the CYP2C19 gain-of-function allele CYP2C19*17.[44] Patients treated with clopidogrel who were carriers of CYP219*17 ($n=1977$) had a higher frequency of PLATO-defined major bleeding than those with no loss- or gain-of-function alleles ($n=1856$) (11.9% vs. 9.5%, $p=0.022$), but no impact on ischemic events was observed.

In the CURE study, 5059 ACS patients were genotyped for CYP2C19*17 gain-of-function alleles.[47] Carriers of CYP2C19*17 derived a greater benefit from clopidogrel compared to placebo than noncarriers (among carriers, HR 0.55, 95% CI 0.42–0.73; among noncarriers, HR 0.85, 0.68–1.05; $p=0.02$ for interaction). There was no evidence that carriage of CYP2C19*17 was associated with a greater incidence of major bleeding in clopidogrel-treated patients compared to placebo ($p$ for heterogeneity$=0.64$).

### 3.6.5 Systematic review and meta-analysis

A meta-analysis has analyzed the association between the gain-of-function CYP2C19*17 polymorphism and the risk of cardiovascular events and bleeding in 11 studies published prior to August 2011.[60] Meta-analysis of three studies showed that carriers of CYP2C19*17 alleles were less likely to be classified as having high on-treatment platelet reactivity than noncarriers (37.9% vs. 50.8%, OR 0.60, 95% CI 0.45–0.79, $p=0.0003$). Analysis of eight studies (12,416 patients) showed that carriers of CYP2C19*17 alleles had 12% less risk of adverse cardiovascular events than noncarriers (9.8% vs. 11.1%; OR 0.86, 95% CI 0.76–0.97, $p=0.01$). In four studies of patients with coronary artery disease that assessed bleeding, carriers of CYP2C19*17 alleles had 23% more risk of bleeding compared with noncarriers (8.0% vs. 6.5%, OR 1.25, 95% CI 1.07–1.47).

## 3.7. *ABCB1* genotype

The *ABCB1* gene (also known as the *MDR1* gene, located on chromosome 7) encodes for the intestinal efflux transporter P-gp. Intestinal absorption of clopidogrel and consequent clopidogrel bioavailability has been shown to be associated with *ABCB1* genotype.[21] It has been hypothesized by multiple studies that the substitution of a T for a C at nucleotide 3435 of the *ABCB1* gene (*ABCB1* C3435T) is associated with reduced absorption of clopidogrel. The prevalence of heterozygotes for the *ABCB1* C3435T allele (*ABCB1* 3435 CT) is approximately 50–70%.[42,61,62] The prevalence of patients homozygous for the T allele (*ABCB1* 3435 TT) appears to be more variable depending on the study, at approximately 2–30%.[42,61]

### *3.7.1* ABCB1 *genotype and PK/PD studies*

The literature regarding the effect of *ABCB1* genotype on PK/PD response to clopidogrel is somewhat contradictory and it is unclear whether *ABCB1* genotype influences exposure to clopidogrel and subsequent platelet reactivity. There was no association of *ABCB1* genotype with AUC or $C_{max}$ of the clopidogrel active metabolite concentration curve in a study of 160 healthy volunteers who were noncarriers for CYP2C19 loss-of-function alleles.[27] *ABCB1* genotype was also not associated with platelet reactivity determined by MEA in a study of 1524 clopidogrel-treated patients who underwent PCI.[61] However, another smaller study of 300 clopidogrel-treated PCI patients found that carriers of *ABCB1* 3435 T alleles had significantly higher platelet reactivity determined by VerifyNow.[63] Interestingly, there was a suggestion that the influence of *ABCB1* genotype on platelet reactivity was greatest at baseline and appeared to diminish over a 6-month period. This was in contrast to CYP2C19 polymorphisms, which appeared to have a consistent effect for a 6-month period. Another smaller study again showed that *ABCB1* polymorphisms did not significantly affect platelet reactivity determined by LTA and VerifyNow in a study of 126 clopidogrel-treated PCI patients.[64] Clopidogrel-treated ACS patients undergoing PCI who were homozygous for the *ABCB1* T allele were more likely to be classified as low responders to clopidogrel, determined by PFA-100 in a study of 98 patients ($p = 0.017$), although the genotype was not associated with adverse cardiovascular events.[65]

### 3.7.2 Key observational studies investigating clinical outcomes

The evidence from observational studies regarding the influence of *ABCB1* genotype on the clinical outcomes of ACS patients is somewhat contradictory. Out of the 2208 patients genotyped for *ABCB1* status in the FAST-MI registry, patients who were homozygous for the *ABCB1* 3435 T allele had a higher risk of adverse cardiovascular events than those who were homozygous for the wild type (15.5% vs. 10.7%, HR 1.72, 95% CI 1.20–2.47).[42] Another study of 928 clopidogrel-treated patients with acute MI did not, however, show an association between *ABCB1* genotype and the occurrence of adverse cardiovascular events.[43]

The role of *ABCB1* genotype in the outcomes of PCI patients is also difficult to interpret. In a study of 1034 clopidogrel-treated PCI patients, *ABCB1* genotype was not associated with adverse cardiovascular events in univariate analysis.[66]

In the study of 300 PCI patients by Campo *et al.*, carriage of *ABCB1* T allele was a risk factor for adverse cardiovascular events in univariate analysis, but it did not appear to be an independent risk factor.[63]

*ABCB1* genotype has not been shown to influence the incidence of stent thrombosis in observational studies. *ABCB1* genotype was not shown to be associated with occurrence of stent thrombosis in a large case–control study of 176 patients with stent thrombosis[67] or in another smaller case–control study of 66 patients with stent thrombosis.[61]

### 3.7.3 Key clinical trials

The PLATO study is the largest single study to investigate the influence of *ABCB1* genotype on clinical outcomes. Ten thousand two hundred and eighty-five ACS patients were genotyped for the *ABCB1* C3435T polymorphism.[44] *ABCB1* genotype was not shown to have a significant influence on the incidence of the primary endpoint (CV death, MI, or stroke) or PLATO-defined major bleeding associated with clopidogrel treatment ($p$ for interaction = 0.39 and 0.80, respectively). The occurrence of the primary endpoint in clopidogrel-treated patients who were homozygous for the wild-type C allele, heterozygous for the T allele, and homozygous for the T allele was 11.9%, 9.8%, and 10.5%, respectively. The occurrence of PLATO-defined major bleeding in each of these groups was 10.8%, 10.6%, and 10.9%, respectively. In the TRITON-TIMI 38 study, 2932 ACS patients were genotyped for the *ABCB1* C3435T polymorphism. Clopidogrel-treated patients who were homozygous for the T allele had an increased risk of the primary endpoint (cardiovascular death, MI, or stroke) compared to patients who were not homozygous for the mutation

(HR 1.72, 52/414 [12.9%] vs. 80/1057 [7.8%], 95% CI 1.22–2.44, $p=0.0020$) at 15-month follow-up. This increased risk was most marked in the first 30 days (corresponding HR 1.96). Neither the G2677T/A nor the C1236T *ABCB1* variants were associated with an increased risk of the primary endpoint.

A Chinese study of 536 PCI patients by Wang *et al.* has investigated modifying clopidogrel loading dose on the basis of *ABCB1* genotype.[62] Sixty-eight percent of the patients were *ABCB1* wild-type homozygous (CC), 30% were heterozygote (CT), and 2% were homozygote for the T allele (TT). After a 300-mg clopidogrel loading dose, 25% of patients with CC alleles were classified as low responders to clopidogrel as determined by VASP phosphorylation. Patients were more likely to be classified as low responders if they possessed CT or TT alleles (76% and 64%, respectively). This low response to clopidogrel in *ABCB1* T allele carriers was largely overcome by treatment with additional loading doses of clopidogrel. There was no association between *ABCB1* genotype and the incidence of adverse cardiovascular events at 1 month, although this is difficult to interpret since carriers of *ABCB1* T alleles received additional loading doses of clopidogrel if they were determined to be low responders.

### 3.7.4 Meta-analysis

A meta-analysis has investigated the association between *ABCB1* genotype, platelet reactivity, and the occurrence of adverse cardiovascular events in clopidogrel-treated patients in 12 studies published prior to March 2012.[68] Analysis of four studies did not demonstrate an association between carriage of the *ABCB1* T allele and platelet reactivity (although one large positive study by Wang *et al.* was excluded due to "suspicious results"). Overall, analysis of 10 studies also did not show a significant association between carriage of the *ABCB1* T allele and adverse cardiovascular events (OR 1.09, 95% CI 0.77–1.54).

Another meta-analysis has also investigated the association between the *ABCB1* genotype and the occurrence of adverse cardiovascular events in clopidogrel-treated patients in six studies published prior to September 2011.[69] This study also did not show a significant association between carriage of the *ABCB1* T allele and adverse cardiovascular events (OR 1.13, 95% CI 0.78–1.64, $p=0.51$).

These meta-analyses are, however, difficult to interpret, since they are based on a small number of heterogeneous studies with differing patient populations, genotype definitions, and methodologies.

### *3.7.5 Summary of the effect of* ABCB1 *polymorphisms*

*ABCB1* polymorphisms have not definitively been shown to affect exposure to clopidogrel active metabolite, platelet reactivity, or clinical outcomes in clopidogrel-treated patients.

## 3.8. Paraoxonase-1

In 2011, a study somewhat controversially suggested that the esterase PON1, not CYP2C19, was the main determinant of the bioactivation of clopidogrel.[24] *In-vitro* analysis of microsomal expression system of metabolizing enzymes suggested that PON1 was the rate-determining enzyme for the conversion of 2-oxo-clopidogrel to its pharmacologically active thiol metabolite. The authors then followed this with a case-cohort study of patients with stable coronary artery disease who were treated with clopidogrel and underwent PCI. *PON1* genotype was compared in patients who developed stent thrombosis and a randomly selected group of patients who did not develop stent thrombosis. Patients who were *PON1* QQ192 homozygous, as opposed to *PON1* RR192 or QR192 genotypes, were more likely to be in the stent thrombosis group than in the control group (OR 3.6; 95% CI 1.6–7.9; $p=0.003$). They also replicated their results in a prospective cohort study of 1982 ACS patients. *PON1* QQ192 homozygous patients were again more likely to develop stent thrombosis than RR192 homozygous patients (HR 10.20; 95% CI 4.39–71.43; $p<0.001$).

However, subsequent studies have not been able to reproduce these findings. In a subsequent crossover study, post-ACS patients were randomly exposed to a 300- or 900-mg loading dose of clopidogrel.[25] *PON1* Q192 or L55M genotypes were not significantly associated with either the $C_{max}$ or the AUC of clopidogrel active metabolite after either loading dose. Similarly, *PON1* genotype was not associated with platelet reactivity after clopidogrel treatment, as determined by the VerifyNow and LTA. This was in contrast to CYP2C19 genotype, which was significantly associated with the generation of clopidogrel active metabolite and subsequent platelet reactivity. Another similar study also did not show an association of *PON1* genotype with the formation of clopidogrel active metabolite or effect on platelet reactivity.[26] This study also showed that microsomes expressing PON1 were not capable of converting 2-oxo-clopidogrel to clopidogrel active metabolite. There was also no correlation between *PON1* genotype and pharmacokinetic or pharmacodynamic response to clopidogrel. *PON1* genotype was also not found to affect platelet reactivity determined by MEA in a cohort study of 1524 patients undergoing PCI or the occurrence of stent thrombosis in a case–control study of 127 patients with stent thrombosis.[28]

## 3.9. Carboxylesterase

After clopidogrel is absorbed, approximately 85% of the prodrug is rapidly hydrolyzed by esterases to become inactivated metabolites and hepatic CES1 is a major catalyst in this process. Certain single nucleotide polymorphisms of *CES1* that confer a loss of function have been identified, resulting in decreased enzymatic activity and attenuation of the hydrolysis of the clopidogrel prodrug and intermediate metabolites into inactive metabolites.[70,71] This therefore results in a greater production of the thiol-containing active metabolite[71] and greater reduction in platelet reactivity.[70]

## 3.10. CYP3A4 genotype

The IVS10 + 12G>A polymorphism of the CYP3A4 gene is associated with less platelet expression of activated GPIIb/IIIa, but similar platelet aggregability, compared to common alleles in patients treated with coronary artery disease treated with clopidogrel.[72] In healthy volunteers exposed to clopidogrel, CYP3A4 genotype was not shown to affect ADP-induced platelet reactivity determined by LTA.[73]

## 3.11. *PEAR1*

A genome-wide association study of 565 healthy Caucasian volunteers treated with dual antiplatelet therapy has suggested that *PEAR1* may be a major determinant of on-treatment platelet reactivity.[74] Genotyping suggested that *PEAR1* genotype accounted for approximately 5% of the variation in collagen-induced platelet aggregation as determined by LTA. As *PEAR1* genotype affects collagen-induced, but not ADP-induced, platelet aggregation, this may suggest that *PEAR1* genotype affects response to aspirin rather than clopidogrel. These findings were replicated in studies of patients with coronary artery disease, who were found to have a higher risk of cardiovascular events or death if they were carriers of *PEAR1* A-alleles compared to noncarriers. Further work is required to pursue these preliminary findings.

# 4. PRASUGREL

## 4.1. Introduction

Prasugrel is a third-generation thienopyridine, which reduces platelet reactivity to a greater degree than clopidogrel.[75] This is largely due to a greater generation of prasugrel active metabolite, as clopidogrel and prasugrel active metabolites are chemically similar and equipotent.[76] Prasugrel reduces the incidence of adverse cardiovascular events in ACS patients who are planned to undergo PCI, compared to clopidogrel.[77]

## 4.2. Metabolism of prasugrel

The metabolism of prasugrel requires fewer hepatic steps than that of clopidogrel, and hence it is thought to be less affected by variation in function of CYP enzymes. Prasugrel is initially de-esterified to an intermediate metabolite by carboxylesterase 2 during absorption and is converted into its active metabolite in one step by CYP3A4, CYP2B6, and, to a lesser degree, CYP2C19 and CYP2C9.[78,79]

## 4.3. Effect of CYP2C19 genotype on the effect of prasugrel

### *4.3.1 Observational studies investigating the effect of CYP2C19 genotype on PK/PD response to prasugrel in healthy volunteers*

In studies in which healthy volunteers receive prasugrel, CYP2C19 loss-of-function alleles have not been shown to have a significant effect on the pharmacokinetics of prasugrel[80] or the pharmacodynamic response determined by LTA,[32,33,80] VerifyNow,[33] or VASP phosphorylation.[33] In a study of 238 healthy volunteers who received prasugrel,[80] genetic variations of CYP2C19, CYP2C9, CYP2B6, CYP3A5, or CYP1A2 did not appear to influence exposure to prasugrel active metabolite or platelet reactivity determined by LTA. Another study randomized healthy volunteers to receive prasugrel ($n=71$) or clopidogrel ($n=74$) to determine the effect of CYP2C19 and CYP2C9 genotype on pharmacokinetic and pharmacodynamic response to these drugs.[32] Again, neither CYP2C19 nor CYP2C9 loss-of-function alleles were shown to influence exposure to active metabolite of prasugrel or its effect on platelet reactivity determined by LTA. In contrast, both CYP219 and CYP2C9 were shown to affect response to clopidogrel. One study randomized 90 Chinese healthy volunteers to receive either prasugrel or clopidogrel, in order to investigate the effect of CYP2C19 loss-of-function alleles on pharmacokinetic and pharmacodynamic response.[33] Carriage of CYP2C19 loss-of-function alleles did not significantly affect exposure to prasugrel active metabolite or platelet reactivity determined by LTA, VASP, and VerifyNow.

### 4.3.2 Observational studies investigating the effect of CYP2C19 genotype on PK/PD response to prasugrel in patients with cardiovascular disease

The majority of studies involving prasugrel-treated patients with cardiovascular disease have also shown no association between CYP2C19 genotype and platelet reactivity determined by LTA,[81] MEA,[81] VerifyNow,[35,36,81,82] and VASP phosphorylation.[36,81,83] CYP2C19 genotype did not affect the pharmacokinetic or pharmacodynamic response of 51 prasugrel-treated patients with stable coronary artery disease in a study that randomized patients to receive either clopidogrel or prasugrel.[36] In this study, CYP2C19 genotype did not influence the effect of prasugrel on platelet reactivity determined by VASP phosphorylation and VerifyNow. Only one dataset (published in two papers)[84,85] has shown an effect of CYP2C19 genotype on platelet reactivity in prasugrel-treated patients. This study of 213 prasugrel-treated ACS patients undergoing PCI assessed the effect of CYP2C19*2 and *17 alleles on platelet reactivity determined by VASP phosphorylation. Carriers of the CYP2C19*2 loss-of-function allele had significantly higher platelet reactivity than noncarriers ($p=0.03$). Carriers of the CYP2C19*17 gain-of-function allele had significantly lower platelet reactivity ($p=0.03$).

### 4.3.3 Clinical trials investigating the effect of CYP2C19 genotype on pharmacodynamic effect of prasugrel

CYP2C19 genotype had no influence on the benefit of prasugrel compared to clopidogrel in 56 ACS patients who previously had high platelet reactivity on clopidogrel therapy.[81] After patients had been switched from clopidogrel to prasugrel, CYP2C19 genotype did not affect subsequent platelet reactivity determined by LTA, MEA, VASP phosphorylation, and VerifyNow. Another study randomized clopidogrel-treated post-PCI patients with high platelet reactivity determined by VerifyNow to receive either prasugrel or an increased dose (150 mg daily) of clopidogrel.[35] The reduction in platelet reactivity associated with prasugrel was not affected by carriage of the CYP2C19*2 loss-of-function allele. A study of 187 ACS patients undergoing PCI randomized patients to a treatment strategy based on point-of-care genotyping or standard care.[82] In the genotyping group, carriers of the loss-of-function CYP2C19*2 allele were treated with prasugrel and noncarriers were treated with clopidogrel. The standard treatment group received clopidogrel only. No patients in the genotyping group had high platelet reactivity (PRU >234 determined by VerifyNow) at 7-day follow-up compared to 30% of those given standard treatment ($p=0.0092$). CYP2C19 genotype also did not influence the effect of prasugrel in a crossover study that investigated the effect of switching 32 clopidogrel-treated elective PCI patients with high platelet reactivity to prasugrel.[83] Presence of the CYP2C19*2 loss-of-function allele was not shown to influence platelet reactivity determined by VASP phosphorylation in the prasugrel-treated patients.

### *4.3.4 Key studies investigating the effect of CYP2C19 genotype on clinical outcomes of prasugrel-treated patients*

In the TRITON study, 1466 prasugrel-treated ACS patients were genotyped.[80] None of CYP2C19, CYP2C9, CYP2B6, CYP3A5, or CYP1A2 genotypes were shown to affect the risk of adverse cardiovascular events in prasugrel-treated ACS patients.

In the only study to have shown an influence of CYP2C19 genotype on the pharmacodynamic effect of prasugrel, carriers of the CYP2C19*17 gain-of-function allele also had a higher rate of bleeding complications than non-carriers ($p=0.02$).

### *4.3.5 Summary of the effect of CYP2C19 polymorphisms on prasugrel-treated patients*

The majority of studies have shown that CYP2C19 genotype does not affect exposure to prasugrel active metabolite, pharmacodynamic response assessed by platelet reactivity, or clinical outcomes in prasugrel-treated patients.

## 4.4. *ABCB1*

In a study that included a substudy from TRITON, 1461 prasugrel-treated patients and 321 healthy volunteers were genotyped for *ABCB1* polymorphisms.[46] In prasugrel-treated subjects, *ABCB1* genotype was not significantly associated with platelet reactivity or exposure to prasugrel active metabolite. Similarly, *ABCB1* genotype did not affect the incidence of the primary endpoint or major bleeding in prasugrel-treated patients.

# 5. TICAGRELOR

Ticagrelor is a nucleoside analogue, which is a novel type of $P2Y_{12}$ inhibitor. It has been shown to reduce the incidence of adverse cardiovascular events and mortality in ACS patients compared to clopidogrel.[86]

### 5.1. Influence of genetic polymorphisms on the effect of ticagrelor

Ten thousand two hundred and eighty-five ACS patients were genotyped in the PLATO trial.[44] The occurrence of the primary endpoint (CV death, MI, and stroke) in ticagrelor-treated patients was not significantly different in carriers of CYP2C19 loss-of-function alleles and noncarriers (8.6% vs. 8.8%). The presence of 0, 1, or 2 *ABCB1* T alleles also did not have a significant effect on the occurrence of the primary endpoint (9.5% vs. 8.5% vs. 8.8%, respectively). CYP2C19 loss-of-function alleles and *ABCB1* genotype had no significant effect on the occurrence of major bleeding in ticagrelor-treated patients.

The effect of ticagrelor on platelet reactivity determined by LTA was not significantly influenced by genotypic variation in *P2RY12*, *P2RY1*, and *ITGB3* in a study of 151 ticagrelor-treated patients.[87]

Ticagrelor does not require metabolic activation to have an effect *in vivo*, but it is metabolized into active and inactive metabolites by CYP3A4 and CYP3A5.[88] Therefore, it is recommended to avoid concurrent treatment with strong CYP3A4/5 inhibitors. However, it has not yet been demonstrated whether CYP3A4/5 polymorphisms affect ticagrelor pharmacokinetics or pharmacodynamics.

## 6. GPIIb/IIIa ANTAGONISTS

GPIIb/IIIa antagonists block the final common pathway of platelet aggregation by inhibiting the cross-linking of platelets via fibrinogen. GPIIb/IIIa antagonists are used frequently as an adjunct during PCI.

### 6.1. $Pl^{A2}$ polymorphisms in healthy volunteer studies

As previously discussed in Section 2 of this review, the $Pl^{A2}$ polymorphism affects the GPIIIa subunit of the GPIIb/IIIa receptor. A study of 62 healthy volunteers and 177 patients with stable coronary artery disease did not show an association between $Pl^{A2}$ genotype and the effect of *in-vitro* abciximab, eptifibatide, and tirofiban on platelet reactivity, as determined by LTA with ADP as an agonist.[89] The pharmacodynamic effect of eptifibatide determined by LTA was attenuated in carriers of the $Pl^{A2}$ allele compared to noncarriers in one study of 31 healthy volunteers.[90] However, another study of 35 healthy volunteers did not show that $Pl^{A2}$ influenced the effect of eptifibatide on platelet reactivity determined by LTA.[91] A study of 56 healthy volunteers showed that carriage of one $Pl^{A2}$ allele was associated with a potentiated effect of *in-vitro* abciximab on inhibiting platelet aggregation compared to noncarriers.[12] However, this is difficult to interpret, as carriage of two $Pl^{A2}$ alleles did not actually appear to influence the effect of abciximab compared to noncarriers.

## 6.2. $Pl^{A2}$ polymorphisms in ACS or post-PCI patients

A study of 1014 ACS patients has investigated the association between $Pl^{A2}$ genotype and the relative effects of orbofiban (an oral GPIIb/IIIa antagonist) compared to placebo.[92] In the orbofiban-treated group, carriers of the $Pl^{A2}$ allele had a higher risk of the primary endpoint (death, MI, recurrent ischemia, urgent revascularization, and stroke) compared to noncarriers (RR 1.55, 95% CI 1.03–2.34, $p=0.04$). Carriers of $Pl^{A2}$ also had a significantly lower incidence of bleeding. A study of 292 patients with ST-elevation MI has investigated whether $Pl^{A2}$ polymorphisms influence the relative benefit of PCI plus abciximab compared to thrombolysis.[93] There was no association between $Pl^{A2}$ genotype and measures or infarct size and mortality in abciximab-treated patients. A study of 87 patients undergoing PCI showed that patients who were heterozygous for the $Pl^{A2}$ allele ($Pl^{A1/A2}$) had a lower response to abciximab, demonstrated by an attenuated reduction in platelet reactivity and fewer fibrinogen binding receptors than noncarriers.[94] This was however not seen in $Pl^{A2}$ homozygotes.

### 6.2.1 Summary of the effect of $Pl^{A2}$ polymorphisms

The $Pl^{A2}$ polymorphism may influence the effect of GPIIb/IIIa antagonists, but the evidence is inconclusive regarding the effect of the polymorphisms on pharmacodynamic response and clinical outcomes in GPIIb/IIIa antagonist-treated patients.

# 7. SUMMARY OF THE GENETICS OF ANTIPLATELET DRUG RESPONSES

Genetic variation in response to antiplatelet drugs appears to be most clinically relevant in patients treated with clopidogrel. Loss-of-function polymorphisms of CYP2C19, the gene encoding for the key enzyme in the metabolism of clopidogrel, are associated with reduced formation of the active metabolite of clopidogrel and a lower effect of the drug, resulting in higher on-treatment platelet reactivity. Many observational studies, subgroup analyses of large randomized controlled trials, and meta-analyses have shown that CYP2C19 loss-of-function polymorphisms are associated with an increased incidence of adverse cardiovascular events in patients treated with clopidogrel. This is particularly relevant in high-risk patients, such as those with ACS undergoing PCI. Conversely, gain-of-function polymorphisms of CYP2C19 are associated with greater formation of the active metabolite of clopidogrel and a higher effect of the drug, resulting in lower on-treatment platelet reactivity. Observational studies, subgroup analyses of randomized controlled trials, and meta-analyses have shown that CYP2C19 gain-of-function polymorphisms are associated with an increased incidence of bleeding in patients treated with clopidogrel. The clinical relevance of genetic variation in *ABCB1* and *PON1* in clopidogrel-treated patients has not been established. It is also uncertain whether genetic variation in response to aspirin, ticagrelor, prasugrel, or GPIIb/IIIa inhibitors is clinically relevant. It has not yet been shown whether or not CYP3A4/5 polymorphisms affect ticagrelor pharmacokinetics or pharmacodynamics.

## REFERENCES


1. Hamm CW, Bassand J-P, Agewall S, et al. ESC guidelines for the management of acute coronary syndromes in patients presenting without persistent ST-segment elevation: the Task Force for the management of acute coronary syndromes (ACS) in patients presenting without persistent ST-segment elevation of the European Society of Cardiology (ESC). *Eur Heart J*. 2011;32(23):2999–3054. http://dx.doi.org/10.1093/eurheartj/ehr236.
2. Task Force on the management of ST-segment elevation acute myocardial infarction of the European Society of Cardiology (ESC), Steg PG, James SK, et al. ESC guidelines for the management of acute myocardial infarction in patients presenting with ST-segment elevation. *Eur Heart J*. 2012;33(20):2569–2619. http://dx.doi.org/10.1093/eurheartj/ehs215.
3. O'Gara PT, Kushner FG, Ascheim DD, et al. 2013 ACCF/AHA guideline for the management of ST-elevation myocardial infarction: a report of the American College of Cardiology Foundation/American Heart Association Task Force on Practice Guidelines. *Circulation*. 2013;127(4):e362–e425. http://dx.doi.org/10.1161/CIR.0b013e3182742cf6.
4. Jneid H, Anderson JL, Wright RS, et al. 2012 ACCF/AHA focused update of the guideline for the management of patients with unstable angina/non-ST-elevation myocardial infarction (updating the 2007 guideline and replacing the 2011 focused update): a report of the American College of Cardiology Foundation/American Heart Association Task Force on Practice Guidelines. *J Am Coll Cardiol*. 2012;60(7):645–681. http://dx.doi.org/10.1016/j.jacc.2012.06.004.
5. Storey RF. Variability of response to antiplatelet therapy. *Eur Heart J*. 2008;10(suppl A): A21–A27. http://dx.doi.org/10.1080/09537100210000244402.
6. Bonello L, Mancini J, Pansieri M, et al. Relationship between post-treatment platelet reactivity and ischemic and bleeding events at 1-year follow-up in patients receiving prasugrel. *J Thromb Haemost*. 2012;10(10):1999–2005. http://dx.doi.org/10.1111/j.1538-7836.2012.04875.x.
7. Bonello L, Pansieri M, Mancini J, et al. High on-treatment platelet reactivity after prasugrel loading dose and cardiovascular events after percutaneous coronary intervention

in acute coronary syndromes. *J Am Coll Cardiol.* 2011;58(5):467–473. http://dx.doi.org/10.1016/j.jacc.2011.04.017.
8. Bonello L, Tantry US, Marcucci R, et al. Consensus and future directions on the definition of high on-treatment platelet reactivity to adenosine diphosphate. *J Am Coll Cardiol.* 2010;56(12):919–933. http://dx.doi.org/10.1016/j.jacc.2010.04.047.
9. Antithrombotic Trialists' Collaboration. Collaborative meta-analysis of randomised trials of antiplatelet therapy for prevention of death, myocardial infarction, and stroke in high risk patients. *BMJ.* 2002;324(7329):71–86.
10. Weiss EJ, Bray PF, Tayback M, et al. A polymorphism of a platelet glycoprotein receptor as an inherited risk factor for coronary thrombosis. *N Engl J Med.* 1996;334(17):1090–1094. http://dx.doi.org/10.1056/NEJM199604253341703.
11. Ridker PM, Hennekens CH, Schmitz C, Stampfer MJ, Lindpaintner K. PIA1/A2 polymorphism of platelet glycoprotein IIIa and risks of myocardial infarction, stroke, and venous thrombosis. *Lancet.* 1997;349(9049):385–388. http://dx.doi.org/10.1016/S0140-6736(97)80010-4.
12. Michelson AD, Furman MI, Goldschmidt-Clermont P, et al. Platelet GP IIIa PlA polymorphisms display different sensitivities to agonists. *Circulation.* 2000;101(9):1013–1018. http://dx.doi.org/10.1161/01.CIR.101.9.1013.
13. Goodman T, Ferro A, Sharma P. Pharmacogenetics of aspirin resistance: a comprehensive systematic review. *Br J Clin Pharmacol.* 2008;66(2):222–232. http://dx.doi.org/10.1111/j.1365-2125.2008.03183.x.
14. Halushka M. Genetic variation in cyclooxygenase 1: effects on response to aspirin. *Clin Pharmacol Ther.* 2003;73(1):122–130. http://dx.doi.org/10.1067/mcp.2003.1.
15. Clappers N, van Oijen MGH, Sundaresan S, et al. The C50T polymorphism of the cyclooxygenase-1 gene and the risk of thrombotic events during low-dose aspirin therapy. *Thromb Haemost.* 2008;100:70–75. http://dx.doi.org/10.1160/TH08-03-0172.
16. Oijen MGH, Laheij RJF, Koetsier M, et al. Effect of a specific cyclooxygenase-gene polymorphism (A-842G/C50T) on the occurrence of peptic ulcer hemorrhage. *Dig Dis Sci.* 2006;51(12):2348–2352. http://dx.doi.org/10.1007/s10620-006-9475-8.
17. The Clopidogrel in Unstable Angina to Prevent Recurrent Events Trial Investigators. Effects of clopidogrel in addition to aspirin in patients with acute coronary syndromes without ST-segment elevation. *N Engl J Med.* 2001;345(7):494–502. http://dx.doi.org/10.1056/NEJMoa010746.
18. Sabatine MS, Cannon CP, Gibson CM, et al. Addition of clopidogrel to aspirin and fibrinolytic therapy for myocardial infarction with ST-segment elevation. *N Engl J Med.* 2005;352(12):1179–1189. http://dx.doi.org/10.1056/NEJMoa050522.
19. Steinhubl SR, Berger PB, Mann JT, et al. Early and sustained dual oral antiplatelet therapy following percutaneous coronary intervention: a randomized controlled trial. *JAMA.* 2002;288(19):2411–2420.
20. Mehta SR, Yusuf S, Peters RJ, et al. Effects of pretreatment with clopidogrel and aspirin followed by long-term therapy in patients undergoing percutaneous coronary intervention: the PCI-CURE study. *Lancet.* 2001;358(9281):527–533.
21. Taubert DD, Beckerath von NN, Grimberg GG, et al. Impact of P-glycoprotein on clopidogrel absorption. *Clin Pharmacol Ther.* 2006;80(5):486–501. http://dx.doi.org/10.1016/j.clpt.2006.07.007.
22. Kazui M, Nishiya Y, Ishizuka T, et al. Identification of the human cytochrome p450 enzymes involved in the two oxidative steps in the bioactivation of clopidogrel to its pharmacologically active metabolite. *Drug Metab Dispos.* 2009;38(1):92–99. http://dx.doi.org/10.1124/dmd.109.029132.
23. Thomas MR, Storey RF. Optimal management of antiplatelet therapy and proton pump inhibition following percutaneous coronary intervention. *Curr Treat Options Cardiovasc Med.* 2011;14(1):24–38. http://dx.doi.org/10.1007/s11936-011-0157-2.

24. Bouman HJ, Schömig E, Van Werkum JW, et al. Paraoxonase-1 is a major determinant of clopidogrel efficacy. *Nat Med.* 2010;17(1):110–116. http://dx.doi.org/10.1038/nm.2281.
25. Hulot JS, Collet JP, Cayla G, et al. CYP2C19 but not PON1 genetic variants influence clopidogrel pharmacokinetics, pharmacodynamics, and clinical efficacy in post-myocardial infarction patients. *Circ Cardiovasc Interv.* 2011;4(5):422–428. http://dx.doi.org/10.1161/CIRCINTERVENTIONS.111.963025.
26. Gong IY, Crown N, Suen CM, et al. Clarifying the importance of CYP2C19 and PON1 in the mechanism of clopidogrel bioactivation and in vivo antiplatelet response. *Eur Heart J.* 2012;33(22):2856–2864. http://dx.doi.org/10.1093/eurheartj/ehs042.
27. Frelinger III AL, Bhatt DL, Lee RD, et al. Clopidogrel pharmacokinetics and pharmacodynamics vary widely despite exclusion or control of polymorphisms (CYP2C19, ABCB1, PON1), noncompliance, diet, smoking, co-medications (including proton pump inhibitors), and pre-existent variability in platelet function. *J Am Coll Cardiol.* 2013;61(8):872–879. http://dx.doi.org/10.1016/j.jacc.2012.11.040.
28. Sibbing D, Koch W, Massberg S, et al. No association of paraoxonase-1 Q192R genotypes with platelet response to clopidogrel and risk of stent thrombosis after coronary stenting. *Eur Heart J.* 2011;32(13):1605–1613. http://dx.doi.org/10.1093/eurheartj/ehr155.
29. Gurbel PA. Clopidogrel for coronary stenting: response variability, drug resistance, and the effect of pretreatment platelet reactivity. *Circulation.* 2003;107(23):2908–2913. http://dx.doi.org/10.1161/01.CIR.0000072771.11429.83.
30. Rudez G, Bouman HJ, Van Werkum JW, et al. Common variation in the platelet receptor P2RY12 gene is associated with residual on-clopidogrel platelet reactivity in patients undergoing elective percutaneous coronary interventions. *Circ Cardiovasc Genet.* 2009;2(5):515–521. http://dx.doi.org/10.1161/CIRCGENETICS.109.861799.
31. Jeong YH, Tantry US, Kim IS, et al. Effect of CYP2C19*2 and *3 loss-of-function alleles on platelet reactivity and adverse clinical events in East Asian acute myocardial infarction survivors treated with clopidogrel and aspirin. *Circ Cardiovasc Interv.* 2011;4(6):585–594. http://dx.doi.org/10.1161/CIRCINTERVENTIONS.111.962555.
32. Brandt JT, Close SL, Iturria SJ, et al. Common polymorphisms of CYP2C19 and CYP2C9 affect the pharmacokinetic and pharmacodynamic response to clopidogrel but not prasugrel. *J Thromb Haemost.* 2007;5(12):2429–2436. http://dx.doi.org/10.1111/j.1538-7836.2007.02775.x.
33. Kelly RP, Close SL, Farid NA, et al. Pharmacokinetics and pharmacodynamics following maintenance doses of prasugrel and clopidogrel in Chinese carriers of CYP2C19 variants. *Br J Clin Pharmacol.* 2011;73(1):93–105. http://dx.doi.org/10.1111/j.1365-2125.2011.04049.x.
34. Simon T, Bhatt DL, Bergougnan L, et al. Genetic polymorphisms and the impact of a higher clopidogrel dose regimen on active metabolite exposure and antiplatelet response in healthy subjects. *Clin Pharmacol Ther.* 2011;90(2):287–295. http://dx.doi.org/10.1038/clpt.2011.127.
35. Alexopoulos D, Dimitropoulos G, Davlouros P, et al. Prasugrel overcomes high on-clopidogrel platelet reactivity post-stenting more effectively than high-dose (150-mg) clopidogrel. *JACC Cardiovasc Interv.* 2011;4(4):403–410. http://dx.doi.org/10.1016/j.jcin.2010.12.011.
36. Varenhorst C, James S, Erlinge D, et al. Genetic variation of CYP2C19 affects both pharmacokinetic and pharmacodynamic responses to clopidogrel but not prasugrel in aspirin-treated patients with coronary artery disease. *Eur Heart J.* 2009;30(14):1744–1752. http://dx.doi.org/10.1093/eurheartj/ehp157.
37. Judge HM, Patil SB, Buckland RJ, Jakubowski JA, Storey RF. Potentiation of clopidogrel active metabolite formation by rifampicin leads to greater P2Y12 receptor

blockade and inhibition of platelet aggregation after clopidogrel. *J Thromb Haemost.* 2010;8(8):1820–1827. http://dx.doi.org/10.1111/j.1538-7836.2010.03925.x.
38. Bouman HJ, Parlak E, van Werkum JW, et al. Which platelet function test is suitable to monitor clopidogrel responsiveness? A pharmacokinetic analysis on the active metabolite of clopidogrel. *J Thromb Haemost.* 2009;8(3):482–488. http://dx.doi.org/10.1111/j.1538-7836.2009.03733.x.
39. Breet NJ. Comparison of platelet function tests in predicting clinical outcome in patients undergoing coronary stent implantation. *JAMA.* 2010;303(8):754. http://dx.doi.org/10.1001/jama.2010.181.
40. Hulot JS. Cytochrome P450 2C19 loss-of-function polymorphism is a major determinant of clopidogrel responsiveness in healthy subjects. *Blood.* 2006;108(7):2244–2247. http://dx.doi.org/10.1182/blood-2006-04-013052.
41. Sibbing D, Stegherr J, Latz W, et al. Cytochrome P450 2C19 loss-of-function polymorphism and stent thrombosis following percutaneous coronary intervention. *Eur Heart J.* 2008;30(8):916–922. http://dx.doi.org/10.1093/eurheartj/ehp041.
42. Simon T, Verstuyft C, Mary-Krause M, et al. Genetic determinants of response to clopidogrel and cardiovascular events. *N Engl J Med.* 2009;360(4):363–375. http://dx.doi.org/10.1056/NEJMoa0808227.
43. Tiroch KA, Sibbing D, Koch W, et al. Protective effect of the CYP2C19 *17 polymorphism with increased activation of clopidogrel on cardiovascular events. *Am Heart J.* 2010;160(3):506–512. http://dx.doi.org/10.1016/j.ahj.2010.06.039.
44. Wallentin L, James S, Storey RF, et al. Effect of CYP2C19 and ABCB1 single nucleotide polymorphisms on outcomes of treatment with ticagrelor versus clopidogrel for acute coronary syndromes: a genetic substudy of the PLATO trial. *Lancet.* 2010;376(9749):1320–1328. http://dx.doi.org/10.1016/S0140-6736(10)61274-3.
45. Holmes MV. CYP2C19 genotype, clopidogrel metabolism, platelet function, and cardiovascular events: a systematic review and meta-analysis. *JAMA.* 2011;306(24):2704. http://dx.doi.org/10.1001/jama.2011.1880.
46. Giusti B, Abbate R. Response to antiplatelet treatment: from genes to outcome. *Lancet.* 2010;376(9749):1278–1281. http://dx.doi.org/10.1016/S0140-6736(10)61313-X.
47. Paré G, Mehta SR, Yusuf S, et al. Effects of CYP2C19 genotype on outcomes of clopidogrel treatment. *N Engl J Med.* 2010;363(18):1704–1714. http://dx.doi.org/10.1056/NEJMoa1008410.
48. Shah AP, Beohar N, Youngquist S, Josephson G, Rosborough JP, Niemann JT. Myocardial ischemia and infarction. *J Am Coll Cardiol.* 2008;51(10):A178–A235. http://dx.doi.org/10.1016/j.jacc.2008.02.007.
49. Mega JL. Reduced-function CYP2C19 genotype and risk of adverse clinical outcomes among patients treated with clopidogrel predominantly for PCIA meta-analysis. *JAMA.* 2010;304(16):1821. http://dx.doi.org/10.1001/jama.2010.1543.
50. Hulot J-S, Collet J-P, Silvain J, et al. Cardiovascular risk in clopidogrel-treated patients according to cytochrome P450 2C19*2 loss-of-function allele or proton pump inhibitor coadministration. *J Am Coll Cardiol.* 2010;56(2):134–143. http://dx.doi.org/10.1016/j.jacc.2009.12.071.
51. Montalescot G, Hulot J-S, Collet J-P. Stent thrombosis: who's guilty? *Eur Heart J.* 2009;30(22):2685–2688.
52. Jin B, Ni HC, Shen W, Li J, Shi HM, Li Y. Cytochrome P450 2C19 polymorphism is associated with poor clinical outcomes in coronary artery disease patients treated with clopidogrel. *Mol Biol Rep.* 2010;38(3):1697–1702. http://dx.doi.org/10.1007/s11033-010-0282-0.
53. Sofi F, Giusti B, Marcucci R, Gori AM, Abbate R, Gensini GF. Cytochrome P450 2C19*2 polymorphism and cardiovascular recurrences in patients taking clopidogrel: a meta-analysis. *Pharmacogenomics J.* 2010;11(3):199–206. http://dx.doi.org/10.1038/tpj.2010.21.

54. Bauer T, Bouman HJ, van Werkum JW, Ford NF, ten Berg JM, Taubert D. Impact of CYP2C19 variant genotypes on clinical efficacy of antiplatelet treatment with clopidogrel: systematic review and meta-analysis. *BMJ*. 2011;343:d4588.
55. Sorich MJ, Polasek TM, Wiese MD. Systematic review and meta-analysis of the association between cytochrome P450 2C19 genotype and bleeding. *Thromb Haemost*. 2012;108(1):199–200. http://dx.doi.org/10.1160/TH12-02-0095.
56. Li-Wan-Po A, Girard T, Farndon P, Cooley C, Lithgow J. Pharmacogenetics of CYP2C19: functional and clinical implications of a new variant CYP2C19* 17. *Br J Clin Pharmacol*. 2010;69(3):222–230.
57. Sibbing D, Gebhard D, Koch W, et al. Isolated and interactive impact of common CYP2C19 genetic variants on the antiplatelet effect of chronic clopidogrel therapy. *J Thromb Haemost*. 2010;8(8):1685–1693. http://dx.doi.org/10.1111/j.1538-7836.2010.03921.x.
58. Frere C, Cuisset T, Gaborit B, Alessi MC, Hulot JS. The CYP2C19*17 allele is associated with better platelet response to clopidogrel in patients admitted for non-ST acute coronary syndrome. *J Thromb Haemost*. 2009;7(8):1409–1411. http://dx.doi.org/10.1111/j.1538-7836.2009.03500.x.
59. Sibbing D, Koch W, Gebhard D, et al. Cytochrome 2C19*17 allelic variant, platelet aggregation, bleeding events, and stent thrombosis in clopidogrel-treated patients with coronary stent placement. *Circulation*. 2010;121(4):512–518. http://dx.doi.org/10.1161/CIRCULATIONAHA.109.885194.
60. Li Y, Tang HL, Hu YF, Xie HG. The gain-of-function variant allele CYP2C19* 17: a double-edged sword between thrombosis and bleeding in clopidogrel-treated patients. *J Thromb Haemost*. 2012;10(2):199–206.
61. Jaitner J, Morath T, Byrne RA, et al. No association of ABCB1 C3435T genotype with clopidogrel response or risk of stent thrombosis in patients undergoing coronary stenting. *Circ Cardiovasc Interv*. 2012;5(1):82–88. http://dx.doi.org/10.1161/CIRCINTERVENTIONS.111.965400.
62. Wang X-D, Zhang D-F, Liu X-B, et al. Modified clopidogrel loading dose according to platelet reactivity monitoring in patients carrying ABCB1 variant alleles in patients with clopidogrel resistance. *Euro J Intern Med*. 2012;23(1):48–53. http://dx.doi.org/10.1016/j.ejim.2011.07.016.
63. Campo G, Parrinello G, Ferraresi P, et al. Prospective evaluation of on-clopidogrel platelet reactivity over time in patients treated with percutaneous coronary intervention. *J Am Coll Cardiol*. 2011;57(25):2474–2483. http://dx.doi.org/10.1016/j.jacc.2010.12.047.
64. Jeong Y-H, Kim I-S, Park Y, et al. Carriage of cytochrome 2C19 Polymorphism is associated with risk of high post-treatment platelet reactivity on high maintenance-dose clopidogrel of 150 mg/day. *JACC Cardiovasc Interv*. 2010;3(7):731–741. http://dx.doi.org/10.1016/j.jcin.2010.05.007.
65. Spiewak M, Malek LA, Kostrzewa G, et al. Influence of C3435T multidrug resistance gene-1 (MDR-1) polymorphism on platelet reactivity and prognosis in patients with acute coronary syndromes. *Kardiol Pol*. 2009;67(8):827–836.
66. Carlquist JF, Knight S, Horne BD, et al. Cardiovascular risk among patients on clopidogrel anti-platelet therapy after placement of drug-eluting stents is modified by genetic variants in both the CYP2C19 and ABCB1 genes. *Thromb Haemost*. 2013;109(4):744–754. http://dx.doi.org/10.1160/TH12-05-0336.
67. Harmsze AM, van Werkum JW, Ten Berg JM, et al. CYP2C19*2 and CYP2C9*3 alleles are associated with stent thrombosis: a case-control study. *Eur Heart J*. 2010;31(24):3046–3053. http://dx.doi.org/10.1093/eurheartj/ehq321.
68. Su J, Xu J, Li X, et al. ABCB1 C3435T polymorphism and response to clopidogrel treatment in coronary artery disease (CAD) patients: a meta-analysis. *PLoS One*. 2012;7(10):e46366. http://dx.doi.org/10.1371/journal.pone.0046366.t011.

69. Luo M, Li J, Xu X, Sun X, Sheng W. ABCB1 C3435T polymorphism and risk of adverse clinical events in clopidogrel treated patients: a meta-analysis. *Thromb Res.* 2012;129(6):754–759. http://dx.doi.org/10.1016/j.thromres.2011.12.003.
70. Lewis JP, Horenstein RB, Ryan K, et al. The functional G143E variant of carboxylesterase 1 is associated with increased clopidogrel active metabolite levels and greater clopidogrel response. *Pharmacogenet Genomics.* 2013;23(1):1–8. http://dx.doi.org/10.1097/FPC.0b013e32835aa8a2.
71. Zhu H-J, Wang X, Gawronski BE, Brinda BJ, Angiolillo DJ, Markowitz JS. Carboxylesterase 1 as a determinant of clopidogrel metabolism and activation. *J Pharmacol Exp Ther.* 2013;344(3):665–672. http://dx.doi.org/10.1124/jpet.112.201640.
72. Angiolillo DJ. Contribution of gene sequence variations of the hepatic cytochrome P450 3A4 enzyme to variability in individual responsiveness to clopidogrel. *Arterioscler Thromb Vasc Biol.* 2006;26(8):1895–1900. http://dx.doi.org/10.1161/01.ATV.0000223867.25324.1a.
73. Fontana P, Hulot JS, De Moerloose P, GAUSSEM P. Influence of CYP2C19 and CYP3A4 gene polymorphisms on clopidogrel responsiveness in healthy subjects. *J Thromb Haemost.* 2007;5(10):2153–2155. http://dx.doi.org/10.1111/j.1538-7836.2007.02722.x.
74. Lewis JP, Ryan K, O'Connell JR, et al. Genetic variation in PEAR1 is associated with platelet aggregation and cardiovascular outcomes. *Circ Cardiovasc Genet.* 2013;6(2):184–192. http://dx.doi.org/10.1161/CIRCGENETICS.111.964627.
75. Michelson AD, Frelinger AL, Braunwald E, et al. Pharmacodynamic assessment of platelet inhibition by prasugrel vs. clopidogrel in the TRITON-TIMI 38 trial. *Eur Heart J.* 2009;30(14):1753–1763.
76. Sugidachi A, Ogawa T, Kurihara A, et al. The greater in vivo antiplatelet effects of prasugrel as compared to clopidogrel reflect more efficient generation of its active metabolite with similar antiplatelet activity to that of clopidogrel's active metabolite. *J Thromb Haemost.* 2007;5(7):1545–1551. http://dx.doi.org/10.1111/j.1538-7836.2007.02598.x.
77. Wiviott SD, Trenk D, Frelinger AL, et al. Prasugrel compared with high loading- and maintenance-dose clopidogrel in patients with planned percutaneous coronary intervention: the Prasugrel in Comparison to Clopidogrel for Inhibition of Platelet Activation and Aggregation Thrombolysis in Myocardial Infarction 44 Trial. *Circulation.* 2007;116(25): 2923–2932. http://dx.doi.org/10.1161/CIRCULATIONAHA.107.740324.
78. Farid NA, Kurihara A, Wrighton SA. Metabolism and disposition of the thienopyridine antiplatelet drugs ticlopidine, clopidogrel, and prasugrel in humans. *J Clin Pharmacol.* 2013;50(2):126–142. http://dx.doi.org/10.1177/0091270009343005.
79. Rehmel JLF. Interactions of two major metabolites of prasugrel, a thienopyridine antiplatelet agent, with the cytochromes P450. *Drug Metab Dispos.* 2006;34(4):600–607. http://dx.doi.org/10.1124/dmd.105.007989.
80. Mega JL, Close SL, Wiviott SD, et al. Cytochrome P450 genetic polymorphisms and the response to prasugrel: relationship to pharmacokinetic, pharmacodynamic, and clinical outcomes. *Circulation.* 2009;119(19):2553–2560. http://dx.doi.org/10.1161/CIRCULATIONAHA.109.851949.
81. Saucedo JF, Angiolillo DJ, DeRaad R, et al. Decrease in high on-treatment platelet reactivity (HPR) prevalence on switching from clopidogrel to prasugrel: insights from the switching anti-platelet (SWAP) study. *Thromb Haemost.* 2013;109(2):347–355. http://dx.doi.org/10.1160/TH12-06-0378.
82. Roberts JD, Wells GA, Le May MR, et al. Point-of-care genetic testing for personalisation of antiplatelet treatment (RAPID GENE): a prospective, randomised, proof-of-concept trial. *Lancet.* 2012;379(9827):1705–1711. http://dx.doi.org/10.1016/S0140-6736(12)60161-5.
83. Sardella G, Calcagno S, Mancone M, et al. Pharmacodynamic effect of switching therapy in patients with high on-treatment platelet reactivity and genotype variation with high

clopidogrel dose versus prasugrel: the RESET GENE Trial. *Circ Cardiovasc Interv.* 2012;5(5):698–704. http://dx.doi.org/10.1161/CIRCINTERVENTIONS.112.972463.
84. Grosdidier C, Quilici J, Loosveld M, et al. Effect of CYP2C19*2 and *17 genetic variants on platelet response to clopidogrel and prasugrel maintenance dose and relation to bleeding complications. *Am J Cardiol.* 2013;111(7):985–990.
85. Cuisset T, Loosveld M, Morange P-E, et al. CYP2C19*2 and *17 alleles have a significant impact on platelet response and bleeding risk in patients treated with prasugrel after acute coronary syndrome. *JACC Cardiovasc Interv.* 2012;5(12):1280–1287. http://dx.doi.org/10.1016/j.jcin.2012.07.015.
86. Wallentin L, Becker RC, Budaj A, et al. Ticagrelor versus clopidogrel in patients with acute coronary syndromes. *N Engl J Med.* 2009;361(11):1045–1057. http://dx.doi.org/10.1056/NEJMoa0904327.
87. Storey RF, Melissa Thornton S, Lawrance R, et al. Ticagrelor yields consistent dose-dependent inhibition of ADP-induced platelet aggregation in patients with atherosclerotic disease regardless of genotypic variations in P2RY12, P2RY1, and ITGB3. *Platelets.* 2009;20(5):341–348. http://dx.doi.org/10.1080/09537100903075324.
88. Zhou D, Andersson TB, Grimm SW. In vitro evaluation of potential drug-drug interactions with ticagrelor: cytochrome P450 reaction phenotyping, inhibition, induction, and differential kinetics. *Drug Metab Dispos.* 2011;39(4):703–710. http://dx.doi.org/10.1124/dmd.110.037143.
89. Weber AA, Jacobs C, Meila D, et al. No evidence for an influence of the human platelet antigen-1 polymorphism on the antiplatelet effects of glycoprotein IIb/IIIa inhibitors. *Pharmacogenetics.* 2002;12(7):581–583.
90. Rozalski M, Watala C. Antagonists of platelet fibrinogen receptor are less effective in carriers of PlA2 polymorphism of β3 integrin. *Eur J Pharmacol.* 2002;454(1):1–8. http://dx.doi.org/10.1016/S0014-2999(02)02391-9.
91. Sirotkina OV, Khaspekova SG, Zabotina AM, Shimanova YV, Mazurov AV. Effects of platelet glycoprotein IIb-IIIa number and glycoprotein IIIa Leu33Pro polymorphism on platelet aggregation and sensitivity to glycoprotein IIb-IIIa antagonists. *Platelets.* 2007;18(7):506–514. http://dx.doi.org/10.1080/09537100701326739.
92. O'Connor FF. Genetic variation in glycoprotein IIb/IIIa (GPIIb/IIIa) as a determinant of the responses to an oral GPIIb/IIIa antagonist in patients with unstable coronary syndromes. *Blood.* 2001;98(12):3256–3260. http://dx.doi.org/10.1182/blood.V98.12.3256.
93. Gorchakova O, Koch W, Mehilli J, et al. PlA polymorphism of the glycoprotein IIIa and efficacy of reperfusion therapy in patients with acute myocardial infarction. *Thromb Haemost.* 2004;91:141–145. http://dx.doi.org/10.1160/TH03-06-0341.
94. Wheeler GL, Braden GA, Bray PF, Marciniak SJ, Mascelli MA, Sane DC. Reduced inhibition by abciximab in platelets with the PlA2 polymorphism. *Am Heart J.* 2002;143(1):76–82.

CHAPTER SEVEN

# The Role of the Transcription Factor *KLF2* in Vascular Development and Disease

**Peter Novodvorsky**[*,†], **Timothy J.A. Chico**[*,†]
[*]MRC Centre for Developmental and Biomedical Genetics, University of Sheffield, Firth Court, Sheffield, United Kingdom
[†]Department of Cardiovascular Science, University of Sheffield, Sheffield, United Kingdom

## Contents



## Abstract

The zinc-finger transcription factor KLF2 transduces the physical forces exerted by blood flow into molecular signals responsible for a wide range of biological responses. Following its initial recognition as a flow-responsive endothelial transcription factor, *KLF2* is now known to be expressed in a range of cell types and to participate in a number of processes during development and disease such as endothelial homeostasis, vasoregulation, vascular growth/remodeling, and inflammation. In this review, we summarize the current understanding about KLF2 with a focus on its effects on vascular biology.

*Progress in Molecular Biology and Translational Science*, Volume 124
ISSN 1877-1173
http://dx.doi.org/10.1016/B978-0-12-386930-2.00007-0

## 1. INTRODUCTION

The endothelium is critically important for maintaining vascular homeostasis and plays an important role in processes such as regulation of vascular tone, inflammation, thrombosis, vasculogenesis/angiogenesis, and atherosclerosis.[1–3] Endothelium forms the interface between circulating blood and the inner layers of the vessel wall and surrounding tissue. Due to this location and function, endothelial phenotype can be affected by both biomechanical or biochemical factors such as blood flow[4] and endogenous[5,6] or pharmacological agents.[7]

Blood flow generates physical forces that act on the vessel wall. These have two major components. Shear stress is a frictional force exerted on the vessel wall with its vector parallel to the direction of flow, whereas tensile stress represents a dilating force on the vessel wall with its vector perpendicular to the direction of flow.[8]

It has been shown that it is not only the absolute magnitude of the shear stress but also, and more importantly, the flow pattern resulting in different shear stress wave forms that is important for the actual effects of shear stress on endothelial phenotype.[9] Atherosclerotic lesions occur in a nonrandom pattern, and their distribution correlates with distinct types of shear stress waveforms. Arterial branch points and areas of major curvatures are associated with turbulent oscillatory flow patterns and significantly higher rates of atherosclerotic lesions when compared to site exposed to pulsatile blood flow with laminar shear stress pattern.[1,9]

## 2. THE KLF FAMILY OF TRANSCRIPTION FACTORS

Krüppel-like factors (KLFs) are zinc-finger transcription factors which were identified by their homology to the *Drosophila melanogaster* protein Krüppel.[10] The mammalian KLF family comprises 17 members so far.[11] Each has three tandem $C_2H_2$ (cysteine–histidine type) zinc fingers in the C terminus of the protein with a consensus amino acid sequence F/Y-X-$\underline{C}$-$X_{2-4}$-$\underline{C}$-$X_3$-F-$X_5$-L-$X_2$-$\underline{H}$-X-R/K-X-$\underline{H}$ (X represents any amino acid, underscored $\underline{C}$ and $\underline{H}$ bind zinc atoms) connected by a characteristic linker T/S-G-E-R/K-P within the DNA-binding domain. These sequences are highly conserved among all members of the KLF family.[12,13] In contrast to this similarity in their DNA-binding domains, the non-DNA-binding N-termini have some common conserved motifs that act as transactivation or repression domains, but generally exhibit much less similarity in their primary structure.[13,14] KLFs act as transcriptional activators and repressors and are expressed in various types of tissues.[15] Their zinc-finger motifs are able to bind to CG-rich sites of general structure CCN CNC CCN, such as CACCC-boxes in various promoters and enhancers.[10,16]

## 2.1. *KLF2*

Our review focuses on KLF2. The majority of experimental work on KLF2 in relation to vascular biology has been performed in human, mouse, and zebrafish. We will therefore briefly detail the *KLF2* genomic structure and expression patterns in these species.

*KLF2* was firstly characterized in 1995 through its homology with *EKLF/KLF1* in mice,[14] followed by its characterization in humans in 1999,[17] and zebrafish in 2001.[13] Due to its high expression in murine lung, it was originally termed lung *KLF(LKLF)*. The murine and human *KLF2* genes are 85% identical, and the mouse and human KLF2 proteins are >90% identical in primary structure.[17] Zebrafish have two *KLF2* paralogues *klf2a* and *klf2b*, reflecting the partial genome duplication of the zebrafish genome after divergence of the teleost lineage during evolution.[18] Zebrafish *klf2a* and *klf2b* are structurally related (61% homology in nucleotide and amino acid sequence) but exhibit different expression patterns and have different functions in developing zebrafish.[13] Zebrafish *klf2a* is considered to be the ortholog of human and murine *KLF2* and like these its expression in the zebrafish vasculature is blood-flow dependent.[13] Zebrafish *klf2a* and human *KLF2* have 49% nucleotide and protein identity. The level of identity is much higher in the region of three tandem zinc fingers—78% for nucleotide identity and 90% for the primary protein structure. *klf2b* expression in zebrafish vasculature has been little studied, although we have detected its expression in the embryonic vasculature by *in situ* hybridization (ISH) (P. N. and T. J. A. C., unpublished data).

Work on murine *Klf2* constructs localized a transcriptional activation domain between amino acids 1–110 and an inhibitory domain between amino acids 111–267. This inhibitory domain directly interacts with the E3 ubiquitin ligase WWP1 which mediates ubiquitination and proteasomal degradation of KLF2.[19,20]

### *2.1.1* KLF2 *expression in human, mouse, and zebrafish*

Northern blot analysis of adult human tissues detected *KLF2* mRNA in heart, lungs, skeletal muscle, pancreas, and placenta (but not brain, liver, or kidney), and vascular tissues were not examined.[17] Human and murine

*KLF2* expression patterns are generally similar.[17] Further experiments confirmed *KLF2* expression in myeloid[21] and lymphoid cells.[22] ISH performed on human vascular tissues from donors of various age (13 months to 76 years) and from different anatomical sites shows that *KLF2* is expressed in all sections of vasculature tested, but the signal strength differs with the predicted patterns of flow and levels of shear stress at these sites; decreased *KLF2* expression is seen at aortic bifurcations.[23] These sites are generally exposed to lower shear stress levels and disturbed flow patterns with atheroprone waveforms. *KLF2* expression levels also negatively correlate with neointimal formation in the human aorta. Within the vessel wall, *KLF2* is selectively expressed in endothelium.[9,23,24] Later work showed that *KLF2* is expressed not only in endothelium of large- and medium-sized vessels but also in the hepatic, duodenal, or the glomerular microvasculature.[25–27]

Northern blot analysis of adult murine tissues shows highest *Klf2* expression in lungs, with some expression detected in the heart, spleen, thymus, skeletal muscle, white and brown adipose tissue, and testes.[14,28,29] The site and level of *Klf2* expression in murine embryonic endothelial cells correspond to the predicted pattern of shear forces in the developing vasculature and remain shear stress dependent in adult murine endothelium.[23,30] *Klf2* expression is detectable in endothelial cells from embryonic day 8.5 (E8.5). With onset of pulsatile blood flow between E8.5 and E10.5, *Klf2* expression rises sharply in endothelial and endocardial cells—a pattern corresponding to elevated shear. *Klf2* expression becomes detectable only on the flow sides of the developing heart valves at E14.5, by E18.5 also in the endocardium lining the intraventricular papillary muscles.[30]

During zebrafish development, at approximately 70% epiboly, *klf2a* can be detected in the ventral, animal portion of the epiblast.[13] In later stages of epiboly *klf2a* expression extends vegetally. At 24 hours post fertilisation (hpf), *klf2a* expression is detectable in the anus, in small clusters of superficial cells lateral to the most posterior notochord, in the cells closely associated with the axial vessels (pronephric ducts), in head vessels, and in the heart. This expression persists until 48 hpf, when a faint signal is also detected in the distal margin of the caudal fin (the tail) and in the mesenchymal interior of the pectoral fin buds.[13,31]

*klf2a* expression in embryonic zebrafish vasculature is like human and mouse *KLF2* blood flow dependent. Vascular expression of *klf2a* can be detected in trunk vasculature from around 36 hpf and is still present at 48 hpf (embryonic zebrafish heart begins to contract at 24–26 hpf[32]). While endothelial *klf2a* is clearly detectable in wild-type embryos, it is lost in *troponin t2* morphant zebrafish which have a noncontractile heart and experience no blood flow.[9,31,33]

## 3. MECHANOTRANSDUCTION AND *KLF2* EXPRESSION

The term *mechanotransduction* refers to multiple mechanisms by which cells convert mechanical forces applied at the cell surface into alteration of gene transcriptional and subsequent changes in cellular signaling.[34] Mechanotransduction is not yet fully understood but comprises multiple components localized in cellular membranes or in the cytosol to activate several cell signaling cascades that each interact. The final cellular response is based on the character of the mechanical stimulus. Mechanical forces acting on the apical surface of an endothelial cell are transmitted through 3D changes of microtubular cytoskeleton to cell–cell and cell–matrix junctions consisting of multiple protein complexes that transduce signals from mechanical forces further into the cell.[35] One such complex consisting of platelet endothelial cell adhesion molecule, vascular endothelial cell cadherin, and vascular endothelial growth factor receptor 2 (VEGFR2) leads subsequently to conformational activation of integrins that mediate both the alignment of endothelial cells under laminar flow conditions and the activation of the proinflammatory NF-κB pathway.[36–38]

Primary cilia are likely to play an accessory and signal-amplifying role in endothelial mechanotransduction. Primary nonmotile cilia are only present in ECs exposed to low and disturbed flow patterns and ECs exposed to laminar flow shed off primary cilia within a timescale of several hours.[39] Primary cilia increase EC response to laminar shear stress measured by induction of *KLF2* which is significantly higher in ciliated ECs versus nonciliated ECs.[40] Consistent with the central role of cytoskeleton in mechanotransduction, ECs chemically depleted of microtubules (Colchicine) or ECs where the microtubular network is stabilized (Taxol/Paclitaxel) show decreased or increased *KLF2* induction by shear stress.[40] Ultrastructurally, two ciliary proteins POLYCYSTIN-1 and POLARIS are necessary for correct shear sensing by primary cilia that results in changes of intracellular calcium and NO levels within the range of several seconds.[41]

Other components that could serve as direct shear stress sensors include the glycocalyx, adhesion molecules such as integrins, cell membrane proteins (receptor tyrosine kinases such as VEGRF2 or G protein coupled receptors), caveolae, and ion channels. All these "primary" sensors are able to transduce the signal further either biochemically through cytoplasm or mechanistically through the cytoskeleton.[42,43] Immediate further steps include phosphorylation of various proteins that activate secondary signaling pathways.[44] Of particular interest in view of *KLF2* are the mitogen-activated protein kinases (MAPKs) that comprise four signaling routes: ERK1/2, ERK5, Jun $NH_2$-terminal kinase (JNK), and p38. MAPKs play essential roles in regulating multiple cellular processes and are responsible for transducing extracellular signals into the cells.[45] ERK5 signaling pathway lies upstream of flow-mediated induction of *KLF2* in endothelium as detailed further.

Finally, a recent theory sees endothelial mechanotransduction as a two-step process: the first, immediate step includes ciliary bending with subsequent intracellular calcium increase and release of NO, endothelin, and other vasoactive substances. In the later prolonged response, cytoskeletal deformations lead to changes in gene expression levels such as *KLF2* which then orchestrate endothelial adaptation to mechanical force.[46]

At the time that *KLF2* was found to be regulated by flow, it was the first endothelial transcription factor reported to have this characteristic.[24] In human umbilical vein endothelial cells (HUVECs) exposed to 24 h of laminar flow (25 dyne/cm$^2$), *KLF2* was 1 of 12 genes (from 18,000) identified to have at least fivefold increased expression compared to static culture. Following this, HUVECs were exposed to unidirectional pulsatile flow ($19 \pm 2$ dyne/cm$^2$) for another 7 days and only 3 of these 12 genes sustained at least fivefold increased expression, including *KLF2*. The fact that pulsatile flow resulted in an additional threefold increase in *KLF2* expression indicates that not only the absolute size of shear stress but also the flow pattern determines *KLF2* expression levels.[24] Further experiments *in vitro* confirmed that regulation of *KLF2* depends on the pattern of flow. HUVECs exposed to 24 h of pulsatile shear with large net forward direction ($12 \pm 4$ dyne/cm$^2$) exhibit a significant increase in *KLF2* expression. HUVECs exposed to oscillatory, atheroprone flow (1Hz $0.5 \pm 4$ dyne/cm$^2$) with low shear stress magnitude and little net forward direction show only a transient increase, followed by a continuous suppression, of *KLF2* expression.[47]

These *in vitro* findings have been confirmed by *in vivo* experiments in several species. *KLF2* levels in human adult vasculature correspond to local shear stress patterns as mentioned above. Similar findings are reported in mice implanted with carotid artery collars[23] and in zebrafish embryos where preventing or stopping blood flow causes significant *klf2a* downregulation.[9] *Klf2* is highly expressed in rat aorta and on the medial aspect of the celiac artery (areas exposed to high levels of laminar shear stress), but is low on the lateral aspect of celiac artery at the branching point with aorta with atheroprone flow patterns[47] Taken together, these findings confirm blood flow-dependent regulation of *KLF2* in all models and species examined.

## 3.1. Flow-dependent regulation of *KLF2*

Figure 7.1 summarizes the mechanisms of regulation of *KLF2* transcription. The critical region required for shear stress-induced expression of *KLF2* lies −157 to −95 bp upstream from the transcription start site and is highly conserved across species.[48] This region contains a single consensus myocyte enhancer factor 2 (MEF2)-binding site.[49] MEF2 binding to this site is not significantly affected by exposure to flow. MEF2 transactivation through

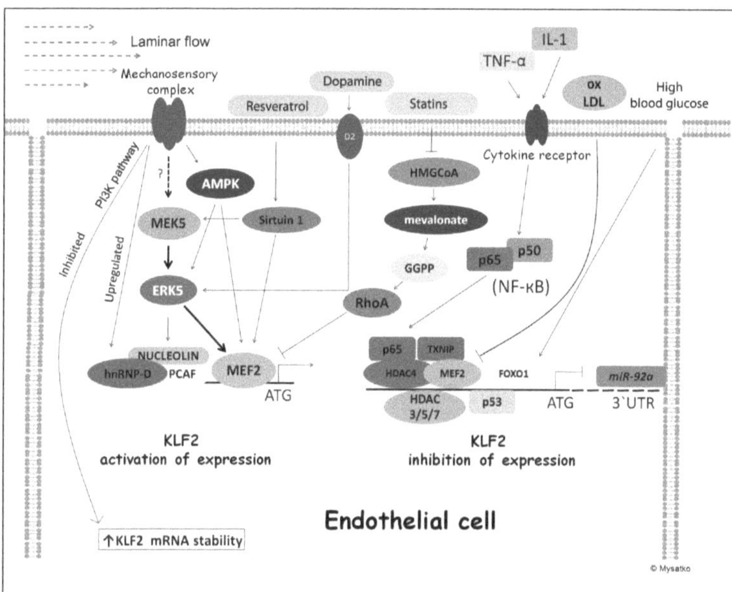


**Figure 7.1** *Molecular mechanisms involved in regulation of endothelial KLF2 expression.* Schematic diagram showing factors that stimulate (left side of diagram) and suppress (right side of diagram) endothelial *KLF2* expression. Abbreviations: KLF2, Krüppel-like factor 2; MEK5, mitogen-activated protein kinase kinase 5; ERK5, extracellular signal-regulated kinase 5; MEF2, myocyte enhancer factor 2; AMPK, AMP-activated protein kinase; PCAF, p300/CBP-associated factor; hnRNP-D, heterogeneous nuclear ribonucleoprotein D; HMG-CoA, 3-hydroxy-3-methylglutaryl-coenzyme A reductase; GGPP, geranylgeranyl pyrophosphate; RhoA, Ras homolog gene family member A; TNF-α, tumor necrosis factor alpha; IL-1, interleukin-1; oxLDL, oxidized low-density lipoprotein; TXNIP, thioredoxin-interacting protein; FOXO1, forkhead box protein O1; HDAC3/4/5/7, histone deacetylase 3/4/5/7; *miR-92a*, microRNA-92a. (See the color plate.)

its phosphorylation is however critical for shear stress-induced *KLF2* upregulation and is mediated via its upstream mitogen-activated protein kinase ERK5 (extracellular signal-regulated kinase 5).[9] ERK5 is in turn specifically upregulated by its activating kinase MEK5 (also known as MAPK5).[9] Another factor necessary for activation of the MEK5/ERK5/MEF2 pathway is AMP-activated protein kinase (AMPK). AMPK lies upstream of MEK5/ERK5/MEF2, and its activation is critical for shear stress-induced phosphorylation of ERK5 and MEF2.[50]

Other cofactors involved in shear stress-mediated *KLF2* regulation include p300/CBP-associated factor (PCAF), heterogeneous nuclear ribonucleoprotein D (hnRNP-D), and nucleolin.[51,52] PCAF and hnRNP-D induce *KLF2* promoter chromatin remodeling via acetylation of histones H3 and H4 with resulting promotion of *KLF2* expression. PCAF and hnRNP-D act through a phosphatidylinositol 3-kinase (PI3K)-dependent, but Akt-independent pathway.[51]

Under static conditions or disturbed flow, histone deacetylase 5 (HDAC5) is bound to MEF2 on the *KLF2* promoter and inhibits MEF2 transcriptional activity. Sufficiently high laminar shear stress (12 dyne/cm$^2$) phosphorylates HDAC5 in a $Ca^{2+}$/calmodulin-dependent manner resulting in its dissociation form MEF2, allowing increased flow-dependent *KLF2* transcription.[53] Oscillatory shear stress promotes expression of class I and II HDAC (HDAC3/5/7) and their association with MEF2 via PI3K/Akt pathway, thus leading to downregulation of *KLF2* expression, whereas pulsatile shear stress induces phosphorylation-dependent class II HDAC (HDAC5/7) nuclear export as described above leading to induction of *KLF2* expression.[54]

Endothelial thioredoxin-interacting protein (*TXNIP*) is downregulated by steady laminar flow which leads to upregulation of thioredoxin activity and subsequent inhibition of EC inflammatory response to tumor necrosis factor alpha (TNF-α).[55] Conversely, *TXNIP* is upregulated by disturbed flow and this promotes endothelial–leukocyte adhesion. TXNIP binds to shear responsive region of *KLF2* promoter (−157 to −78 bp) where it forms a part of a transcriptional repressing complex and inhibits *KLF2* expression.[56]

*microRNAs* (*miRs*) are small (18–24 bp long) single-stranded noncoding RNAs that bind to the 3′-untranslated region (3′-UTR) of their target mRNAs and thus regulate gene expression at the posttranscriptional level, either via translational inhibition or by degradation of mRNAs.[57–59] *miRs* play a role in flow-mediated regulation of *KLF2* expression as evidenced by the fact that knockdown of the critical component of *miRs* biosynthesis pathway Dicer in HUVECs increases *KLF2* expression. 3′-UTR region of *KLF2* contains an *miR-92a*-binding site, and it was shown that laminar flow downregulates *miR-92a* and induces *KLF2* expression, while *miR-92a* overexpression decreases expression of *KLF2* (Fig. 7.1).[60]

# 4. NON-FLOW-DEPENDENT REGULATION OF *KLF2*

## 4.1. Factors stimulating *KLF2* expression

Statins are lipid-lowering drugs widely used in clinical practice. It has been shown that their beneficial effects exceed those expected from lipid-lowering alone.[61] Several statins induce *KLF2* expression in a dose-dependent manner, and KLF2 mediates the induction of endothelial NO synthase (*eNOS*), thrombomodulin (*TM*), and heme oxygenase-1 (*HO-1*) by statins.[7,62]

Statins are inhibitors of 3-hydroxy-3-methylglutaryl-coenzyme A (HMG-CoA) reductase, a crucial enzyme in cholesterol synthesis. As a result of this, cells are depleted of mevalonate which is a precursor of isoprenoid intermediates such as farnesyl pyrophosphate and geranylgeranyl pyrophosphate (GGPP). Isoprenoids function as membrane anchors for various proteins, for example, small GTPases from the Rho family. One of them, Ras homolog gene family member A (RhoA), undergoes posttranslational modification by geranylgeranylation and is then able to inhibit *KLF2* expression.[62] Taken together, statins inhibit the mevalonate pathway and thus production of functionally active RhoA and the absence of RhoA has a positive effect on *KLF2* expression[62] (Fig. 7.1).

Prolonged shear stress induces KLF2-mediated expression of *eNOS* and *TM* compared to statins because shear stress is able to stabilize *KLF2* mRNA via inhibition of PI3K. Specific inhibition of PI3K by LY294002 results in higher stability and thus higher levels of *KLF2* mRNA.[63] This negative role of PI3K pathway contradicts the described positive effect of PI3K pathway on shear-specific nuclear binding of several factors that promote *KLF2* transcription mentioned elsewhere in this review. Shear stress and statins combined have an additive, not synergistic effect on *KLF2*, *eNOS*, and *TM* expression in human endothelial cells.[64]

Statins exert their pleiotropic effects via KLF2 not only in endothelial cells but also in macrophages and lymphocytes. Human macrophages treated with simvastatin exhibit upregulation of *KLF2* expression and downregulation of several proinflammatory chemokines such as monocyte chemotactic protein-1 (MCP-1), macrophage inflammatory proteins-1a and β, interleukin-2 receptor-β, lymphotoxin β, VCAM-1, and TF.[65] Human T-cells treated with simvastatin show significantly increased *KLF2* expression with decreased IFN-γ secretion and diminished T-cell proliferation.[66]

Several lines of evidence have therefore suggested that some clinical effects of statins occur via upregulation of KLF2. Addition of simvastatin to the cold storage solution in which explanted healthy or steatotic rat livers are stored maintains *Klf2* expression with resulting prevention of endothelial dysfunction.[67,68] Atorvastatin increases *Klf2* and decreases *Par-1* receptor expression in aortae of $ApoE^{-/-}$ mice in keeping with its anti-inflammatory effects.[69]

Resveratrol, a polyphenol produced naturally in some plants and present in red wine[70] is a potent inducer of *KLF2* in HUVECs. This induction is mediated via $NAD^{+}$-dependent deacetylase Sirtuin-1 which in turn activates MEK5 and MEF2[71] (Fig. 7.1).

Recently, the structurally and functionally abnormal tumor vasculature[72–75] was found to lack sympathetic innervation and dopamine regulation.[76] Exogenous administration of dopamine acting through D2 receptors can normalize abnormal tumor vessel morphology and leakiness resulting in improved blood flow and reduced tumor hypoxia. This effect may be explained by dopamine's ability to upregulate *KLF2* expression in tumor EC and HUVECs through upregulation of ERK5 (Fig. 7.1). Dopamine also directly upregulates *angiopoietin-1* expression in tumor endothelial pericytes. This could be of therapeutic relevance because administration of DA increases concentration of chemotherapeutic agents in tumor tissues.[76]

Rapamycin (sirolimus) is one of the drugs commonly eluted from coronary stents to reduce restenosis. The effect of rapamycin on the endothelium is complex. Its potentially prothrombotic profile[77–79] might be counteracted by its ability to increase *KLF2* expression in HUVECs.[80] The authors of this study postulate that *KLF2* might be a downstream target of the PI3K/AKT/mTOR pathway. This is based on the fact that rapamycin inhibits the PI3K/AKT/mTOR pathway by inhibition of the mammalian target of rapamycin (mTOR)[81] and there have been reports showing that specific inhibition of PI3K by LY2940002 results in stabilization of *KLF2* mRNA.[63] This is however in conflict with studies, suggesting that a shear stress-dependent binding of several cofactors necessary for *KLF2* transcription is PI3K dependent.[51]

Even more surprisingly, the same group of researchers who initially suggested positive effect of rapamycin on *KLF2* expression in HUVECs[80] soon after published another work suggesting the opposite effect of rapamycin on murine *Klf2* mRNA and protein expression *in vivo*. Here, the rapamycin-induced inhibition of *Klf2* expression in mice resulted in a significantly shorter time to $FeCl_3$-induced murine carotid artery thrombotic occlusion.[82]

## 4.2. Factors inhibiting *KLF2* expression

Several inflammatory cytokines potently inhibit *KLF2* expression. Exposure of HUVECs to IL-1β results in a 4.7-fold decrease in *KLF2* expression,[83] while this is reduced by treatment with TNF-α.[84] TNF-α-mediated repression of *KLF2* occurs via the NF-κB pathway, but does not depend on direct NF-κB binding to the *KLF2* promoter. The p65 subunit of NF-κB cooperates with histone deacetylase 4 (HDAC4) to bind to the *KLF2* promoter and inhibit MEF2-mediated induction of *KLF2* (Fig. 7.1).[49]

Normal and oxidized low-density lipoprotein (LDL) particles potently inhibit *KLF2* expression *in vitro* in HUVECs[85,86] (Fig 7.1). LDL particles exert their inhibitory effect on *KLF2* expression epigenetically via stimulation of DNA methyltransferase 1-induced CpG dinucleotide methylation. Methylated CpG islands decrease MEF2 occupancy of the *KLF2* promoter and promote assembly of a transcriptional repressor complex consisting of methyl-CpG-binding protein 2 (MeCP2) and histone methyltransferase enhancer of zeste homolog 2 (EZH2).[86]

High glucose levels (35 mmol/L) suppress *KLF2* and *eNOS* expression levels in HUVECs and in carotid arteries of diabetic rats.[87] This suppression of *KLF2* is mediated by the forkhead box protein O1 (FOXO1) that directly binds to the *KLF2* promoter (Fig. 7.1). This could represent one possible mechanism of endothelial dysfunction in diabetics.[87] Atorvastatin inhibits the negative effects of FOXO1 on *KLF2* and *eNOS* and restores *KLF2* and *eNOS* expression in HUVECs incubated in high glucose. Similar findings were observed *in vivo* in diabetic rats. Mechanistically, atorvastatin deactivates FOXO1 by its phosphorylation resulting in its translocalization from nucleus into cytoplasm.[88] Interestingly, active FOXO1 completely prevents *KLF2* induction by atorvastatin.[88]

The adaptor protein p66shc promotes cellular oxidative stress[89] and has proapoptotic,[90] proatherogenic,[91] and proangiogenic[92] effects. p66shc downregulates *MEF2A* expression resulting in downregulation of *KLF2* and *TM* mRNA and protein. Conversely, p66shc knockdown increases *KLF2* and *TM* mRNA and protein levels and decreases hydrogen peroxide levels in HUVECs.[93]

p53 is a tumor suppressor gene which plays a crucial role in regulating the cell cycle, DNA, repair and apoptosis.[94] p53 inhibits *KLF2* expression in HUVECs through recruiting HDAC and binding to its p53 response element in *KLF2* promoter (Fig. 7.1) with resulting H3 histone hypoacetylation and subsequent transcriptional repression of *KLF2*.[95]

## 5. FUNCTIONS OF KLF2

Below we summarize the existing literature on the functions of KLF2 in the vasculature (Fig. 7.2). Because of the large number of studies and our increasingly complex understanding of how KLF2 influences vascular homeostasis, we have divided this into sections, although, as will be apparent, much overlap exists.

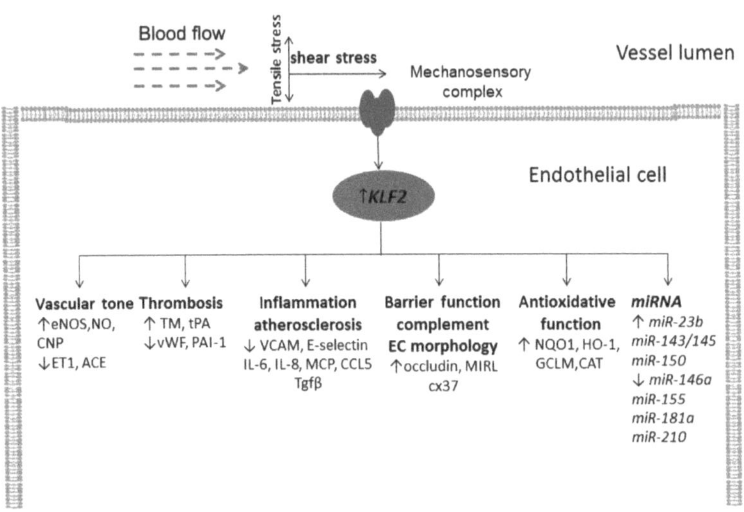


**Figure 7.2** *Effects of* KLF2 *on endothelial homeostasis.* Laminar and pulsatile blood flow induces *KLF2* expression in endothelial cells. KLF2 regulates expression of important endothelial genes involved in endothelial homeostasis. Abbreviations: KLF2, Krüppel-like factor 2; eNOS, endothelial nitric oxide synthase; NO, nitric oxide; CNP, C-natriuretic peptide; ET1, endothelin-1; ACE, angiotensin-converting enzyme; TM, thrombomodulin; tPA, tissue plasminogen activator; vWF, von Willebrand factor; PAI-1, plasminogen activator inhibitor 1; VCAM, vascular cell adhesion molecule; Il-6 and IL-8, interleukin 6 and 8; MCP, monocyte chemotactic protein; CCL5, chemokine (C–C motif) ligand 5; Tgfβ, transforming growth factor β; MIRL, membrane inhibitor of active lysis, also known as CD59; cx37, connexion 37; NQO1, NAD(P)H dehydrogenase quinone 1; HO-1, heme oxygenase 1; GCLM, glutamate–cysteine ligase modifier subunit; CAT, catalase. (See the color plate.)

## 5.1. Maintenance of endothelial homeostasis

KLF2 plays a critical role in flow-mediated upregulation of multiple genes in endothelium. Fifteen percent of all genes known to be upregulated by flow (109 of 716) are KLF2 dependent.[9] Later data suggest that KLF2 together with nuclear factor erythroid2-related factor 2 (Nrf2) controls expression of about 70% of shear stress-induced endothelial genes.[96] KLF2 modulates expression of genes critical in regulating vascular tone, hemostasis/thrombosis, and inflammation and regulates endothelial barrier function and its antioxidative capacity (Fig. 7.2). KLF2 acts as a molecular switch between healthy inactivated and activated atheroprone phenotype with direct effects on pathogenesis of atherosclerosis.[9,97] Hemizygous-deficient *Klf2*$^{+/-}$ mice in *ApoE*$^{-/-}$ background exhibit increased diet-induced atherosclerosis,[97] and levels of *Klf2* expression in vulnerable aortic plaques are significantly decreased when compared with the Klf2 levels in stable plaques of *ApoE*$^{-/-}$ mice.[69] KLF2 has distinct roles in developmental and adult vasculogenesis and angiogenesis and in heart formation detailed in separate sections.

### *5.1.1 Vasoregulation*

Healthy endothelium is able to effectively regulate vascular tone. ECs produce several vasodilators such as nitric oxide (NO) from eNOS, C-natriuretic peptide (CNP) and adrenomedullin, or vasoconstrictive factors such as endothelin-1 (ET1) and angiotensin-converting enzyme (ACE).[98–100] KLF2 potently induces expression of *eNOS* and *CNP*, inhibits expression of endothelin-1 and *ACE*, and surprisingly also inhibits expression of adrenomedullin.[23,83] KLF2 directly binds to the *eNOS* promoter to increase its transcriptional activity and downregulates caveolin-1, a negative regulator of eNOS activity.[9,101]

eNOS and NO have multiple effects on endothelium. Apart from being a vasodilator, NO is anti-inflammatory via inhibition of endothelial adhesion molecules ICAM-1 and VCAM-1[102,103] and antithrombotic.[104] Flow-mediated induction of eNOS through KLF2 therefore represents an important mechanism by which shear stress mediates its effects on endothelium.

### *5.1.2 Thromboprotection*

KLF2 regulates expression of key endothelial thrombotic factors and generally has an antithrombotic effect. Overexpression of *KLF2* in HUVECs significantly increases expression and activity of antithrombotic *TM* and mildly increased expression of tissue plasminogen activator (*tPA*). TM is a key cofactor in the thrombin-induced activation of the anticoagulant protein C. KLF2 directly binds to the *TM* promoter to increase its expression. *KLF2* overexpression inhibits expression of prothrombotic factors plasminogen activator inhibitor (*PAI-1*) and von Willebrand factor (*vWF*).[84] *KLF2* overexpression inhibits cytokine-mediated induction of tissue factor (TF) and PAI-1 and prevents TM inhibition.[84] *In vitro* assays show that perfusion of native human blood over HUVECs overexpressing *KLF2* significantly increases clotting time,[84] while *KLF2* knockdown has the opposite effect, with reduction of antithrombotic gene expression and induction of procoagulant factors.[84]

Interestingly, in contrast, another group observed approximately twofold induction of *vWF* mRNA and protein levels following lentiviral *KLF2* overexpression in HUVECs. After stimulation with thrombin and forskolin, both inducers of vWF release, a corresponding twofold increase of vWF protein amounts was detected.[105] Additionally, several changes were observed in Weibel–Palade bodies, the storage organelles for vWF, and other bioactive compounds. Weibel–Palade bodies in HUVECs overexpressing *KLF2* are shorter, have lower content of proinflammatory angiopoietin-2 and interleukin-8 (IL-8), but are more equally distributed with larger average number per cell across the population of HUVECs when compared to controls.[105,106]

KLF2 is involved in pathogenesis of the antiphospholipid syndrome (APS) characterized by production of antiphospholipid antibodies, particularly against β2-glycoprotein 1 (β2GP1) resulting in thrombosis and recurrent miscarriage.[107] β2GP1 together with anti-β2GP1 causes endothelial cell activation that is central to APS pathogenesis.[108,109] β2GP1/anti-β2GP1 inhibit, via yet unknown mechanisms, expression of *KLF2* and *KLF4*. In the presence of low levels of KLF2 and KLF4, CBP/p300, a transcriptional coactivator of both *NF-κB* and *KLFs*, gets preferentially bound to *NF-κB*. This results in increased NF-κB signaling and endothelial activation.[110]

### 5.1.3 Inflammation

Adenovirally overexpressed *KLF2* in HUVECs inhibits IL-1β, TNFα, and lipopolysaccharide (LPS)-mediated induction of adhesion molecules VCAM-1 and E-selectin.[9] As a consequence, T-cell attachment and rolling are markedly attenuated. KLF2 also inhibits IL-1β-mediated production of inflammatory cytokines.[9] KLF2 exerts anti-inflammatory functions through interactions with several members of the NF-κB pathway and subsequent decrease of NF-κB activity.[83] Thrombin also mediates its proinflammatory effects via the NF-κB pathway. KLF2 inhibits expression of thrombin receptor protease-activated receptor-1 (*PAR-1*) and consequently thrombin-mediated NF-κB nuclear accumulation and DNA binding.[111]

Decreased expression of *KLF2* as a result of proinflammatory signaling activation (TNF-α through NF-κB) causes decreased levels of an anti-inflammatory factor bone morphogenic protein endothelial precursor cell-derived regulator (BMPER). Low levels of BMPER enable increased expression of proinflammatory bone morphogenic protein 2 (*BMP2*) resulting in decreased *eNOS* and increased *ICAM-1* and *VCAM* expression levels and induced leukocyte adhesion and extravasation. Overexpression of *KLF2* in HUVECs increases *BMPER* expression and prevents TNF-α-induced *BMPER* downregulation, providing another line of evidence for the anti-inflammatory effects of KLF2 on endothelium.[112]

Anti-inflammatory effects of prolonged shear stress have been linked to inhibition of MAPK pathways.[113] JNK is an upstream kinase that activates both activator protein-1 (AP-1) components activating transcription factor 2 (ATF2) and c-Jun. AP-1 together with NF-κB and other co-activators like CBP/p300 forms a transcriptional complex that potently induces expression of proinflammatory genes in endothelium in response to inflammatory cytokines and is therefore considered as proinflammatory and proatherogenic.[114] *ATF2* is constitutively expressed in human endothelium but is found in high amounts in its active phosphorylated form in endothelial cells overlying early atherosclerotic lesions. The observation that prolonged shear stress inhibits ATF2 function by nuclear exclusion of phosphorylated ATF2 that is mediated by KLF2[115] was later expanded by the same group when they showed that the anti-inflammatory effects of KLF2 are mediated through actin cytoskeleton changes and require several days to reach full effect.[116] In this model, KLF2 inhibits phosphorylation of focal adhesion kinase involved in actin cytoskeleton regulation[117] and in phosphorylation of JNK (together with actin filaments).[118] As a result, JNK is not activated via phosphorylation and cannot activate AP-1, ATF2, and c-Jun. The same effect could be observed following disruption of actin cytoskeleton by cytochalasin D, confirming the link between KLF2 and inhibition of the JNK pathway via the actin cytoskeleton.[116]

Regulation of *KLF2* expression may differ at different developmental stages and may also depend on the animal model used. This is shown by the interplay between shear stress, KLF2, and transforming growth factor β (Tgfβ) signaling in human nonembryonic endothelial cells such as HUVECs or human aortic endothelial cells (HAoECs) and murine embryonic endothelial cells on the other hand.[119,120]

Tgfβ signaling in nonembryonic endothelial cells is considered to have proatherogenic effects because Tgfβ induces PAI-1, MCP-1, and endothelial oxidized LDL receptor. Shear stress increases *KLF2* expression in these cells, and KLF2 subsequently inhibits Tgfβ signaling in two separate ways. First, KLF2 induces Smad7 that inhibits phosphorylation of Smad2 and transcriptional activity of Smad3/4 which are all members of Tgfβ signaling pathway acting through the Tgfβ receptor Alk5. Second, KLF2 inhibits another Tgfβ signaling pathway by inhibiting phosphorylation of one of the AP-1 components c-Jun.[119]

On the other hand, in murine embryonic endothelial cells, increased shear stress activates Tgfβ/Alk5 signaling. Alk5 in turn activates the MEK5/ERK5/MEF2 pathway and increases *Klf2* expression.[120]

Thus, in mouse embryonic endothelial cells, shear stress-mediated *Klf2* induction lies downstream of Tgfβ/Alk5/MEK5/ERK5 signaling. In HUVECs and HAoECs, shear stress-mediated *KLF2* induction is MEK5/ERK5/MEF2 dependent and KLF2 is upstream of Tgfβ/Alk5 signaling on which KLF2 has an inhibitory effect.

### 5.1.4 Complement activation

Endothelial exposure to laminar but not oscillatory shear stress leads to significant upregulation of MAC inhibitory protein CD59 (MIRL) on the endothelial surface with resulting decreased complement-mediated lysis of flow-conditioned EC. This effect of shear stress is mediated via ERK5/Klf2 signaling pathway.[121]

### 5.1.5 Endothelial barrier function

*Klf2*$^{+/-}$ mice show increased endothelial leakage in response to an inflammatory stimulus.[84] This corresponds with *in vitro* experiments, showing a protective effect of *KLF2* overexpression in HUVECs on thrombin, $H_2O_2$, and histamine-induced endothelial leakage. KLF2 upregulates expression of a key tight-junction protein occludin and decreases phosphorylation of myosin light chain which is a fundamental event in cell contraction.[122] Protective effects of Klf2 on endothelial barrier function were confirmed independently in a murine stroke model. Transient middle cerebral artery occlusion was induced in *Klf2*$^{-/-}$ mice, *Klf2*-overexpressing mice, and control mice, and subsequently infarct volume and blood barrier function were analyzed. As expected, *Klf2*$^{-/-}$ mice exhibited significantly larger infarct volume and impaired blood barrier function due to decreased expression of occludin. Overexpression of *Klf2* reduced infarct volume and blood barrier function was preserved.[123]

Recent reports suggest that decreased KLF2 might be involved in pathogenesis of Alzheimer's disease (AD).[124] Blood–brain barrier dysfunction and impaired endothelial permeability have been implicated in pathogenesis of AD for some time.[125] *Klf2* mRNA and protein levels in Tg2576 mouse model of AD are significantly reduced due to accumulation of amyloid beta that acts through increased p53 levels that subsequently inhibit *Klf2* expression. Overexpression of *KLF2* in human brain EC completely rescues impaired expression of occludin, in keeping with the reported role of KLF2 in endothelial barrier function.[124] It must be noted however that potent KLF2 inducers statins have failed to decrease incidence of AD in randomized controlled trials in humans.[126]

### 5.1.6 Endothelial morphology and intercellular gap junctions

It has been known that flow and shear influence endothelial actin cytoskeleton rearrangement and cell shape for more than two decades.[127,128] The central role of KLF2 in the cytoskeleton changes has now been elucidated. HUVECs exposed to shear stress ($19 \pm 17$ dyne/cm$^2$ for 4 days) align in the direction of flow. This alignment is abrogated after siRNA-mediated *KLF2* silencing, whereas HUVECs overexpressing *KLF2* exhibit a stretched shape as a result of reorganization of the cytoskeleton with formation of stress fibers even in the absence of flow.[116] HUVECs overexpressing *KLF2* also show increased cell density and decreased migration following wounding.[105]

ECs from healthy arteries or EC cultured under high laminar shear stress conditions exhibit high expression of connexin 37 (*CX37*), and its deletion in *ApoE*-deficient mice accelerates atherosclerosis.[129,130] cx37 is a transmembrane protein that forms a gap junction in the form of two hexameric hemichannels on adjacent cells. Gap junctions represent a specialized intercellular connection that links the cytoplasm of two cells enabling direct communication through the exchange of ions and small metabolites.[131] KLF2 regulates *CX37* expression in response to shear stress or simvastatin by directly binding its promoter. Deletion of *KLF2* decreases passage of small molecules and ions through gap junctions in an EC monolayer, confirming the function of KLF2 in intercellular communication.[132]

### 5.1.7 Oxidative stress

HUVECs overexpressing *KLF2* show upregulation of antioxidant genes such as NAD(P)H dehydrogenase quinone 1 (*NQO1*), *HO-1*, glutamate-cysteine ligase modifier subunit (*GCLM*), and catalase (*CAT*), which are all target genes of Nrf2, one of the main antioxidant transcription factors upregulated by atheroprotective blood flow.[87,105,133] KLF2 promotes Nrf2 nuclear localization and activation necessary for its function and augments Nrf2-mediated protection against oxidative stress.[96]

HO-1 is the rate-limiting enzyme in heme catabolism that catalyzes heme degradation into biliverdin, carbon monoxide, and free iron.[134] Products of this pathway have antioxidant,[135] anti-inflammatory,[136] and anti-apoptotic[137] effects on vasculature. Statins act as potent inducers of HO-1 in a KLF2-dependent manner, and their antioxidant effects have been shown to act through the generation of biliverdin and ferritin following HO-1 activation.[138]

### 5.1.8 microRNA production

HUVECs overexpressing *KLF2* were examined for expression of *miRs* known to play roles in vascular homeostasis. Expression levels of *miR-23b* and *miR-150* (regulatory role in cell migration) were significantly increased. Expression levels of *miR-146a* and *miR-155* (role in inflammation), *miR-181a* (endothelial cell fate), and *miR-210* (angiogenesis and hypoxia response) were on the contrary reduced.[139] Interestingly, expression of *miR-143/145* which has an atheroprotective effect on vascular smooth muscle cells (VSMCs) was upregulated in HUVECs under shear stress and lentiviral overexpression of *KLF2*.[139] A novel mechanism has been described in which shear stress and/or KLF2 increases production of extracellular vesicles (ectosomes) in HUVECs expressing high levels of *miR-143/145* and these are secreted from HUVECs to induce a paracrine atheroprotective VSMC phenotype.[139] This communication between endothelial cells and SMC might explain the effect of endothelial *Klf2* knockdown on VSMCs observed a decade earlier.[140,141]

## 5.2. T-cell and B-cell biology

T- and B-lymphocytes play distinct roles in atherosclerosis.[142,143] This section therefore briefly describes the role of KLF2 in lymphocyte biology especially in regard to vascular pathophysiology. Klf2 plays an important role in maintaining a quiescent, nonactivated T-cell phenotype. *Klf2* expression is developmentally induced in quiescent single-positive (SP) T-cells (either

$CD4^+$ or $CD8^+$), but rapidly decreased following T-cell activation. Additionally, *Klf2*-deficient chimeric mice develop a massive reduction in peripheral single-positive T-cells and show spontaneously activated cell surface phenotype and increased Fas-mediated apoptosis.[28] Klf2 maintains T-cell quiescence at least partially via negative regulation of the proto-oncogene *c-Myc*.[144] Maintenance of a quiescent T-cell phenotype by Klf2 might be physiologically important due to the role of activated $CD8^+$ T-lymphocytes in promotion of vulnerable atherosclerotic plaques in *ApoE*-deficient mice.[145] Additionally, Klf2 is necessary for peripheral T-cell recirculation. *Klf2*-deficient SP T-cells show impaired thymic emigration and deficient T-cell trafficking; the majority of $Klf2^{-/-}$ T-cells are found in the spleen with almost none in the blood or lymph nodes. Klf2 was subsequently shown to regulate thymocyte and T-cell trafficking into peripheral lymph organs by inducing expression of receptors critical for these processes, such as *CD62L*, *β-integrin*, *CCR7* (T-cell trafficking), and *S1P1* (thymocyte emigration).[146]

A particular subset of T-cells, the $CD4^+CD25^+Fox3p^+$ T regulatory cells (Tregs) known to have a role in maintaining immunological tolerance, have inhibitory effects on development and progression of atherosclerosis.[147,148] Tregs are able to upregulate *KLF2* expression in HUVECs previously treated with oxidized LDL in a manner requiring direct cell contact.[85]

*Klf2* knockdown in B cells also leads to decreased expression of trafficking molecules *CD62L* and *β-integrin*, but expression of *S1P1* receptor remains almost unaffected. *Klf2* deficiency also causes impaired B-cell subset differentiation with increased number of marginal zone B cells and massively reduced number of atheroprotective B1 B cells.[149]

The antiatherogenic effects of *KLF2* expression in lymphocytes are supported by the observation that simvastatin significantly increases *KLF2* expression in murine and human T-cells *in vitro* and *in vivo* with resulting decreased IFN-γ secretion and diminished T-cell proliferation.[66]

## 5.3. Monocyte and macrophage biology

*KLF2* is expressed in primary human monocytes, but its expression is reduced by cytokine or LPS activation or differentiation into macrophages. *KLF2* expression is reduced by about 30% in monocytes of patients with extensive atherosclerosis compared to healthy controls.[21] This is significant because monocyte activation and recruitment play important roles in atherosclerosis.[150]

In mice, conditional *Klf2* knockout in myeloid cell lineage in an atheroprone LDL receptor-deficient background ($Ldl^{-/-}$) causes increased atherosclerosis[151] in keeping with increased atherosclerosis seen in $Klf2^{+/-}$ $ApoE^{-/-}$ mice.[97] Myeloid *Klf2* knockout increases monocyte and neutrophil adhesion to endothelial cells with resulting increased accumulation and activity in atherosclerotic plaques,[151] but this did not confirm increased lipid accumulation by macrophages as observed in $Klf2^{+/-}$ $ApoE^{-/-}$ mice.[97]

Overexpression studies show that *KLF2* inhibits LPS-mediated activation of monocytes as evidenced by reduced secretion of cyclooxygenase 2 (COX2), TF, IL-1, Il-8, TNF-α, and MCP-1. Overexpressed *KLF2* also decreases phagocytic activity and surprisingly does not inhibit but rather increases recruitment of monocytes to sites of inflammation. Conversely, *KLF2* knockdown increases monocyte expression of *MCP-1*, *TF*, and *COX2*. KLF2 exerts its anti-inflammatory effects in monocytes by inhibiting the NF-κB and AP-1 signaling pathways.[21] KLF2 does not alter NF-κB or AP-1 expression, nuclear accumulation, or DNA binding, but reduces NF-κB and AP-1 transcriptional activities by interacting with the coactivator PCAF.[21]

Macrophages represent a heterogeneous cell population: M1 macrophages (activated by GM-CSF, TNF, or LPS) represent a proinflammatory phenotype, whereas M2 macrophages are generally anti-inflammatory, contributing to tissue repair but with a higher capacity to accumulate oxidized LDL due to higher number of scavenger receptors.[152–154] *KLF2* expression is higher in M2 than in M1 macrophages. However, after exposure to oxidized LDL, *KLF2* expression decreases in M2 but remains unchanged in M1 macrophages. *KLF2* knockdown in M2 macrophages leads to increased secretion of MCP-1.[155] Overall, the above data suggest that KLF2 plays an anti-inflammatory and antiatherogenic role in monocyte and macrophage biology.

## 5.4. Vasculogenesis and angiogenesis

Homozygous *Klf2*-deficient mice die between E12.5 and 14.5 from intraembryonic and intra-amniotic hemorrhaging associated with normal vascular patterning, but defects in blood vessel morphology—endothelial necrosis, cuboidal VSMCs, abnormally thin tunica media, and aneurysms. $Klf2^{-/-}$ mice exhibit defective VSMCs and pericyte migration to endothelial tubes during vasculogenesis with subsequent loss of their stabilizing and modulatory functions and failure to organize into a compact tunica media.[140,141] There are however conflicting data about the effect of homozygous *Klf2* deletion on murine embryonic development. Another group was able to confirm the stage of lethality in *Klf2*$^{-/-}$ embryos (E11.5–13.5) and reported retarded growth, craniofacial abnormalities, signs of anemia, and abdominal bleeding, but could not detect vessel wall abnormalities.[156] Mouse embryos with conditional endothelial or smooth and cardiac muscle *Klf2*$^{-/-}$ deletion confirm that endothelial *Klf2* deletion is responsible for the embryonic mortality around E14.[30] The cause of death was reported to be high-output cardiac failure caused by loss of smooth muscle tone and vasodilation. The reasoning that *Klf2* deletion was responsible for the observed phenotype was backed up by data obtained from *klf2a* morphant zebrafish embryos. These exhibited increased aortic blood velocity and pericardial oedema.[30]

Vascular endothelial growth factor (VEGF-A) is a key regulator of physiologic and pathologic vasculogenesis and angiogenesis.[157] VEGF-A also promotes vascular permeability, EC migration, and survival, but can also act as a proinflammatory cytokine.[158–161] VEGF-A effects are mediated by its receptor tyrosine kinases, of which VEGFR2 is the most important in endothelium.[162] The following sections describe the relationship of KLF2 to VEGF-A signaling, showing that these two factors act in various and context-dependent manners.

*klf2a* (upregulated by flow) plays a positive role in angiogenesis in zebrafish by inducing expression of an endothelial-specific microRNA *miR-126* which inhibits a VEGF signaling inhibitor *spred-1*. This therefore allows VEGF-A/VEGFR2 (*flk1* or *kdrl* in zebrafish)-mediated angiogenesis. In the absence of flow, *klf2a* and *miR-126* are downregulated and *spred-1* inhibits angiogenesis by inhibiting VEGF-A/VEGFR2 signaling. This pathway thus connects flow with *klf2a* and VEGF-A/VEGFR2 via *mir-126* and suggests a stimulatory role of *klf2a* on VEGF-A/VEGFR2 signaling.[163]

The link between KLF2 and VEGF through *miR-126* has been confirmed in human duodenal vasculature when duodenal biopsies from cirrhotic patients were compared to healthy individuals. Increased duodenal angiogenesis in cirrhotic patients was demonstrated by increased capillary density in duodenal villi and by increased presence of endothelial markers CD31 and CD34.[26] Increased KLF2, *miR-126*, VEGF, and angiopoietin-2 (Ang-2) were detected in duodenal samples from cirrhotic patients, suggesting that KLF2 might play a role in formation of portosystemic collateral vessels as a consequence of increased intrahepatic resistance and increased blood flow.[26]

Similarly, in embryonic endothelial cells of *Xenopus*, expression of *Flk1* is independently activated by Klf2 and by one of the ETS transcription factors

ERG. Klf2 and ERG directly bind to *Flk1* enhancer within the first intron of the *Flk1* gene.[164] Mutation of ERG or Klf-binding sites results in complete or significant reduction of *Flk1* expression, respectively. *Klf2* knockdown inhibits *Flk1* expression and causes significant defects in vasculogenesis. Additionally, Klf and ERG can form a protein complex with synergistic effects on *Flk1* expression in *Xenopus* embryonic endothelial cells.[164]

*Vegf-1* is expressed in adult murine arterial ECs but not venous or capillary endothelium.[165] Shear stress potently activates VEGF signaling in HUVECs on various levels by increasing *VEGF-A* mRNA and protein levels and also increases expression and activation of VEGFR2.[165] Increased VEGF-A/VEGFR2 signaling has, through the induction of *KLF2* expression, protective effects on HUVECs by decreasing apoptosis compared to HUVECs in static conditions.[165] This is contradictory to previous work, suggesting an inhibitory role of *Klf2* on VEGF-A/VEGFR2 (FLK1) signaling in adult endothelial cells.[105,166]

*Klf2* overexpression in a nude mouse model markedly attenuates VEGF-A-mediated angiogenesis and edema. *KLF2* overexpression in HUVECs also potently inhibits VEGF-A mediated endothelial activation. KLF2 inhibits expression of *VEGFR2* by directly competing with Sp1 for a binding site in the *VEGFR2* promoter.[166] Overexpression of *KLF2* inhibits endothelial migration, most likely due to induction of an antimigratory factor semaphorin-3F and to some extent to a less pronounced induction of VEGFR2.[105] These differences in the observed relationship between KLF2 and VEGF signaling may be explained by the different developmental stages and model organisms used (zebrafish and *Xenopus* embryonic angiogenesis vs. human and murine adult angiogenesis).

Consistent with an inhibitory role of *KLF2* in adult angiogenesis is that KLF2 inhibits expression and function of hypoxia-inducible factor 1 (*HIF-1*), which is a central regulator of the hypoxic response and angiogenesis in many cell types. Under hypoxic conditions, *KLF2* overexpressed in HUVECs inhibits (and *KLF2* knockdown increases) expression of HIF-1 target genes such as IL-8, Ang-2, or VEGF and also inhibits endothelial tube formation on Matrigel.[167] KLF2 prevents the hypoxia-mediated accumulation of HIF-1$\alpha$ subunit of HIF and thus prevents HIF-1$\alpha$ translocation into the cell nucleus where it associates with oxygen insensitive HIF-1$\beta$ subunit to form a functional HIF molecule. KLF2 inhibits HIF-1$\alpha$ interaction with its chaperone Hsp90 and thus promotes its proteasomal degradation.[167]

Angiopoietin-1 (Ang-1) and its receptor tyrosine kinase Tie2 are involved in maintaining vascular quiescence and angiogenesis.[168,169] The ability to exert these distinct and opposite functions resides in the fact that in the presence of cell–cell contacts, Ang-1 induces *trans*-association of Tie2 and preferentially activates the Akt pathway, leading to vascular quiescence. In the absence of cell–cell contact, Ang-1 induces Tie2 anchoring to intracellular matrix that preferentially activates the ERK1/2 pathway to promote angiogenesis.[170,171] KLF2 is one of the factors induced by *trans*-associated Tie2 in the presence of cell–cell contacts promoting vascular quiescence. Ang-1/Tie2-mediated KLF2 induction depends on PI3K/Akt pathway which in turn activates transcriptional activity of MEF2.[172]

The Grb2-associated binder family docking proteins (Gab1–Gab3) play crucial roles in transmitting signals that control cell growth, differentiation, and function from multiple receptors.[173] Gab1 has proangiogenic properties and plays a crucial role in postischemic angiogenesis and arteriogenesis in mice. In this pathway, hepatocyte growth factor stimulates Gab1 association with SHP2 (Src homology-2 domain-containing protein tyrosine phosphatase2) and PI3K subunit p85. This complex positively regulates migration, proliferation, and stabilization of ECs via distinct MAPK pathways. Interestingly, Gab1/SHP2 stimulates *KLF2* expression through ERK5 that might be contributing to the stabilizing effects of this signaling cascade on EC.[174] Conversely, *Gab1* conditional endothelial knockout in the $ApoE^{-/-}$ background decreases *Klf2* and *Klf4* expression and increases production of proinflammatory TNF-α, IL-1β, and IL-6, and endothelial expression of *VCAM-1* with resulting accelerated angiotensin 2-mediated atherosclerosis and aortic aneurysm formation.[175]

Proangiogenic cells (PACs), also known as endothelial progenitor cells (EPCs), are bone marrow-derived cells which circulate in the blood stream and are able to take part in angiogenesis.[176] PACs numbers and their neovascularization properties are negatively affected by risk factors for ischemic heart disease such as age, hypertension, or smoking.[177] *KLF2* overexpression in human PACs increases their number by 60% *in vitro* and improves neovascularization abilities of aged murine PACs in an ischemic hind limb model *in vivo*.[178] A particular subset of PACs and the endothelial colony-forming cells have the ability to form *de novo* vessels *in vivo* and react to laminar and pulsatile flow in a similar manner to adult EC, with increased *KLF2* and decreased *ET1* and *VCAM1* expression.[179]

## 5.5. Valvulogenesis

Cardiac valves form in vertebrates from endocardial cushions arising in the atrioventricular (AV) canal. Formation of endocardial cushions involves several steps starting with specification of a subset of endocardial cells in the AV canal and ending in their epithelial-to-mesenchymal transformation (EMT). Endocardial cushions then go on to form the AV complex and after further remodeling become functional valve leaflets.[180] There are several lines of evidence that KLF2 plays an important role in EMT.

Before functional valves develop, anterograde and retrograde (reversing) blood flows exist between atrium and ventricle. Eventually with development of mature valves only anterograde flows become possible. Experiments with alteration of the proportion of retrograde flow to the length of a cardiac cycle (the so-called retrograde flow fraction (RFF)) showed that decreased RFF was linked with severe defects in valve formation in zebrafish.[181] *klf2a* is upregulated by retrograde flow in the AV canal of zebrafish hearts and conversely is downregulated when RFF decreases. Interestingly, *klf2a* knockdown results in valvular defects similar to those observed in zebrafish with decreased RFF, suggesting an important role of *klf2a* and its target genes *notch1b* (zebrafish Notch homolog), *bmp4*, *edn1*, and *nrg1* in blood flow-dependent valvulogenesis.[181]

Another line of evidence that *klf2a* might be involved in correct heart valve formation comes through the identification of a novel zebrafish mutant with defective endocardial cushion formation named *bungee* ($bng^{jh177}$) with a deactivating mutation in *protein kinase 2* (*pkd2*) that leads to impaired phosphorylation of Hdac5. Hdac5 thus remains in its active state bound to the *klf2a* promoter and inhibits *klf2a* expression in the AV region, leading to decreased *notch1b* signaling and defective valve formation.[182]

Interestingly, the role of Klf2 in murine cardiac valve development is strain specific. FVB/N $Klf2^{-/-}$ mice exhibit defective EMT and delayed formation of atrial septum, whereas C57BL/6 $Klf2^{-/-}$ mice experience delayed atrial septal formation but their AV cushions develop normally. FVB/N $Klf2^{-/-}$ mice at E9.5 show multiple layers of dysmorphic endothelial cells lining the AV cushions that fail to undergo the EMT properly resulting in hypocellular AV cushions at E10.5 and abnormal cardiac function. Mechanistically, Klf2 regulates several genes involved in AV cushion development such as *Gata4*, *Tbx5*, *Sox9*, and *Ugdh*.[183]

## 6. SUMMARY

KLF2 orchestrates vascular homeostasis and serves as a key molecular switch between proinflammatory atheroprone and quiescent atheroresistant endothelial phenotype. This characteristic is attributed to its ability to activate atheroprotective and inhibit atherogenic endothelial transcription programs. Additionally, antiatherogenic properties of KLF2 appear to be mediated through its effects on lymphocyte and monocyte/macrophage biology. KLF2 has important roles in embryonic and adult vasculogenesis and angiogenesis under physiological and pathological conditions and plays a key role in embryonic heart valve formation. Recent data show that the previously described pleiotropic effects of statins are mediated through KLF2. Studying the mechanisms of *KLF2* expression and its involvement in vascular physiology is therefore of significant importance in the search for novel therapies in cardiovascular diseases or cancer. It is likely that an ability to modulate *KLF2* expression or function would lead to novel therapeutic strategies for a range of human diseases.

## ACKNOWLEDGMENTS

This work was supported by BHF project grants (09/287/28051 and 12/12/29433) awarded to T. J. A. C. MRC Centre Grant (G0700091) provided a clinical training fellowship to P. N.

## REFERENCES

1. Gimbrone Jr MA. Endothelial dysfunction, hemodynamic forces, and atherosclerosis. *Thromb Haemost.* 1999;82(2):722–726.
2. Ross R. Atherosclerosis—an inflammatory disease. *N Engl J Med.* 1999;340(2): 115–126.
3. Folkman J, Haudenschild C. Angiogenesis in vitro. *Nature.* 1980;288(5791):551–556.
4. Topper JN, Gimbrone Jr MA. Blood flow and vascular gene expression: fluid shear stress as a modulator of endothelial phenotype. *Mol Med Today.* 1999;5(1):40–46.
5. Schleef RR, Bevilacqua MP, Sawdey M, Gimbrone Jr MA, Loskutoff DJ. Cytokine activation of vascular endothelium. Effects on tissue-type plasminogen activator and type 1 plasminogen activator inhibitor. *J Biol Chem.* 1988;263(12):5797–5803.
6. Busse R, Trogisch G, Bassenge E. The role of endothelium in the control of vascular tone. *Basic Res Cardiol.* 1985;80(5):475–490.
7. Parmar KM, Nambudiri V, Dai G, Larman HB, Gimbrone Jr MA, Garcia-Cardena G. Statins exert endothelial atheroprotective effects via the KLF2 transcription factor. *J Biol Chem.* 2005;280(29):26714–26719.
8. White CR, Frangos JA. The shear stress of it all: the cell membrane and mechanochemical transduction. *Philos Trans R Soc Lond B Biol Sci.* 2007;362(1484):1459–1467.
9. Parmar KM, Larman HB, Dai G, et al. Integration of flow-dependent endothelial phenotypes by Kruppel-like factor 2. *J Clin Invest.* 2006;116(1):49–58.
10. Miller IJ, Bieker JJ. A novel, erythroid cell-specific murine transcription factor that binds to the CACCC element and is related to the Kruppel family of nuclear proteins. *Mol Cell Biol.* 1993;13(5):2776–2786.
11. van Vliet J, Crofts LA, Quinlan KG, Czolij R, Perkins AC, Crossley M. Human KLF17 is a new member of the Sp/KLF family of transcription factors. *Genomics.* Apr 2006;87(4):474–482.
12. Ruppert JM, Kinzler KW, Wong AJ, et al. The GLI-Kruppel family of human genes. *Mol Cell Biol.* 1988;8(8):3104–3113.
13. Oates AC, Pratt SJ, Vail B, et al. The zebrafish klf gene family. *Blood.* 2001;98(6):1792–1801.

14. Anderson KP, Kern CB, Crable SC, Lingrel JB. Isolation of a gene encoding a functional zinc finger protein homologous to erythroid Kruppel-like factor: identification of a new multigene family. *Mol Cell Biol.* 1995;15(11):5957–5965.
15. Turner J, Crossley M. Mammalian Kruppel-like transcription factors: more than just a pretty finger. *Trends Biochem Sci.* 1999;24(6):236–240.
16. Klevit RE. Recognition of DNA by Cys2, His2 zinc fingers. *Science.* 1991;253(5026):1367–1393.
17. Wani MA, Conkright MD, Jeffries S, Hughes MJ, Lingrel JB. cDNA isolation, genomic structure, regulation, and chromosomal localization of human lung Kruppel-like factor. *Genomics.* 1999;60(1):78–86.
18. Taylor JS, Braasch I, Frickey T, Meyer A, Van de Peer Y. Genome duplication, a trait shared by 22000 species of ray-finned fish. *Genome Res.* Mar 2003;13(3):382–390.
19. Conkright MD, Wani MA, Lingrel JB. Lung Kruppel-like factor contains an autoinhibitory domain that regulates its transcriptional activation by binding WWP1, an E3 ubiquitin ligase. *J Biol Chem.* 2001;276(31):29299–29306.
20. Zhang X, Srinivasan SV, Lingrel JB. WWP1-dependent ubiquitination and degradation of the lung Kruppel-like factor, KLF2. *Biochem Biophys Res Commun.* 2004;316(1):139–148.
21. Das H, Kumar A, Lin Z, et al. Kruppel-like factor 2 (KLF2) regulates proinflammatory activation of monocytes. *Proc Natl Acad Sci USA.* 2006;103(17):6653–6658.
22. Riley JL, Mao M, Kobayashi S, et al. Modulation of TCR-induced transcriptional profiles by ligation of CD28, ICOS, and CTLA-4 receptors. *Proc Natl Acad Sci USA.* 2002;99(18):11790–11795.
23. Dekker RJ, van Thienen JV, Rohlena J, et al. Endothelial KLF2 links local arterial shear stress levels to the expression of vascular tone-regulating genes. *Am J Pathol.* 2005;167(2):609–618.
24. Dekker RJ, van Soest S, Fontijn RD, et al. Prolonged fluid shear stress induces a distinct set of endothelial cell genes, most specifically lung Kruppel-like factor (KLF2). *Blood.* 2002;100(5):1689–1698.
25. Gracia-Sancho J, Russo L, Garcia-Caldero H, Garcia-Pagan JC, Garcia-Cardena G, Bosch J. Endothelial expression of transcription factor Kruppel-like factor 2 and its vasoprotective target genes in the normal and cirrhotic rat liver. *Gut.* 2011;60(4):517–524.
26. Kobus K, Kopycinska J, Kozlowska-Wiechowska A, et al. Angiogenesis within the duodenum of patients with cirrhosis is modulated by mechanosensitive Kruppel-like factor 2 and microRNA-126. *Liver Int.* 2012;32(8):1222–1232.
27. Slater SC, Ramnath RD, Uttridge K, et al. Chronic exposure to laminar shear stress induces Kruppel-like factor 2 in glomerular endothelial cells and modulates interactions with co-cultured podocytes. *Int J Biochem Cell Biol.* 2012;44(9):1482–1490.
28. Kuo CT, Veselits ML, Leiden JM. LKLF: a transcriptional regulator of single-positive T cell quiescence and survival. *Science.* 1997;277(5334):1986–1990.
29. Banerjee SS, Feinberg MW, Watanabe M, et al. The Kruppel-like factor KLF2 inhibits peroxisome proliferator-activated receptor-gamma expression and adipogenesis. *J Biol Chem.* 2003;278(4):2581–2584.
30. Lee JS, Yu Q, Shin JT, et al. Klf2 is an essential regulator of vascular hemodynamic forces in vivo. *Dev Cell.* 2006;11(6):845–857.
31. Wang L, Zhang P, Wei Y, Gao Y, Patient R, Liu F. A blood flow-dependent klf2a-NO signaling cascade is required for stabilization of hematopoietic stem cell programming in zebrafish embryos. *Blood.* 2011;118(15):4102–4110.
32. Kimmel CB, Ballard WW, Kimmel SR, Ullmann B, Schilling TF. Stages of embryonic development of the zebrafish. *Dev Dyn.* 1995;203(3):253–310.

33. Stainier DY, Fouquet B, Chen JN, et al. Mutations affecting the formation and function of the cardiovascular system in the zebrafish embryo. *Development*. 1996;123:285–292.
34. Mammoto A, Mammoto T, Ingber DE. Mechanosensitive mechanisms in transcriptional regulation. *J Cell Sci*. 2012;125(Pt 13):3061–3073.
35. Davies PF. Overview: temporal and spatial relationships in shear stress-mediated endothelial signalling. *J Vasc Res*. 1997;34(3):208–211.
36. Tzima E, del Pozo MA, Shattil SJ, Chien S, Schwartz MA. Activation of integrins in endothelial cells by fluid shear stress mediates Rho-dependent cytoskeletal alignment. *EMBO J*. 2001;20(17):4639–4647.
37. Tzima E, Del Pozo MA, Kiosses WB, et al. Activation of Rac1 by shear stress in endothelial cells mediates both cytoskeletal reorganization and effects on gene expression. *EMBO J*. 2002;21(24):6791–6800.
38. Tzima E, Irani-Tehrani M, Kiosses WB, et al. A mechanosensory complex that mediates the endothelial cell response to fluid shear stress. *Nature*. 2005;437(7057):426–431.
39. Iomini C, Tejada K, Mo W, Vaananen H, Piperno G. Primary cilia of human endothelial cells disassemble under laminar shear stress. *J Cell Biol*. 2004;164(6):811–817.
40. Hierck BP, Van der Heiden K, Alkemade FE, et al. Primary cilia sensitize endothelial cells for fluid shear stress. *Dev Dyn*. 2008;237(3):725–735.
41. Nauli SM, Kawanabe Y, Kaminski JJ, Pearce WJ, Ingber DE, Zhou J. Endothelial cilia are fluid shear sensors that regulate calcium signaling and nitric oxide production through polycystin-1. *Circulation*. 2008;117(9):1161–1171.
42. Tarbell JM, Weinbaum S, Kamm RD. Cellular fluid mechanics and mechanotransduction. *Ann Biomed Eng*. 2005;33(12):1719–1723.
43. Ando J, Yamamoto K. Vascular mechanobiology: endothelial cell responses to fluid shear stress. *Circ J*. 2009;73(11):1983–1992.
44. Davies PF. Hemodynamic shear stress and the endothelium in cardiovascular pathophysiology. *Nat Clin Pract Cardiovasc Med*. 2009;6(1):16–26.
45. Roberts OL, Holmes K, Muller J, Cross DA, Cross MJ. ERK5 and the regulation of endothelial cell function. *Biochem Soc Trans*. 2009;37(Pt 6):1254–1259.
46. Poelmann RE, Van der Heiden K, Gittenberger-de Groot A, Hierck BP. Deciphering the endothelial shear stress sensor. *Circulation*. 2008;117(9):1124–1126.
47. Wang N, Miao H, Li YS, et al. Shear stress regulation of Kruppel-like factor 2 expression is flow pattern-specific. *Biochem Biophys Res Commun*. 2006;341(4):1244–1251.
48. Huddleson JP, Srinivasan S, Ahmad N, Lingrel JB. Fluid shear stress induces endothelial KLF2 gene expression through a defined promoter region. *Biol Chem*. 2004;385(8):723–729.
49. Kumar A, Lin Z, SenBanerjee S, Jain MK. Tumor necrosis factor alpha-mediated reduction of KLF2 is due to inhibition of MEF2 by NF-kappaB and histone deacetylases. *Mol Cell Biol*. 2005;25(14):5893–5903.
50. Young A, Wu W, Sun W, et al. Flow activation of AMP-activated protein kinase in vascular endothelium leads to Kruppel-like factor 2 expression. *Arterioscler Thromb Vasc Biol*. 2009;29(11):1902–1908.
51. Huddleson JP, Ahmad N, Srinivasan S, Lingrel JB. Induction of KLF2 by fluid shear stress requires a novel promoter element activated by a phosphatidylinositol 3-kinase-dependent chromatin-remodeling pathway. *J Biol Chem*. 2005;280(24):23371–23379.
52. Huddleson JP, Ahmad N, Lingrel JB. Up-regulation of the KLF2 transcription factor by fluid shear stress requires nucleolin. *J Biol Chem*. 2006;281(22):15121–15128.
53. Wang W, Ha CH, Jhun BS, Wong C, Jain MK, Jin ZG. Fluid shear stress stimulates phosphorylation-dependent nuclear export of HDAC5 and mediates expression of KLF2 and eNOS. *Blood*. 2010;115(14):2971–2979.

54. Lee DY, Lee CI, Lin TE, et al. Role of histone deacetylases in transcription factor regulation and cell cycle modulation in endothelial cells in response to disturbed flow. *Proc Natl Acad Sci USA*. 2012;109(6):1967–1972.
55. Yamawaki H, Pan S, Lee RT, Berk BC. Fluid shear stress inhibits vascular inflammation by decreasing thioredoxin-interacting protein in endothelial cells. *J Clin Invest*. 2005;115(3):733–738.
56. Wang XQ, Nigro P, World C, Fujiwara K, Yan C, Berk BC. Thioredoxin interacting protein promotes endothelial cell inflammation in response to disturbed flow by increasing leukocyte adhesion and repressing Kruppel-like factor 2. *Circ Res*. 2012;110(4):560–568.
57. Chekulaeva M, Filipowicz W. Mechanisms of miRNA-mediated post-transcriptional regulation in animal cells. *Curr Opin Cell Biol*. Jun 2009;21(3):452–460.
58. Bartel DP. MicroRNAs: genomics, biogenesis, mechanism, and function. *Cell*. 2004;116(2):281–297.
59. Bartel DP. MicroRNAs: target recognition and regulatory functions. *Cell*. 2009;136(2):215–233.
60. Bonauer A, Carmona G, Iwasaki M, et al. MicroRNA-92a controls angiogenesis and functional recovery of ischemic tissues in mice. *Science*. 2009;324(5935): 1710–1713.
61. Bellosta S, Ferri N, Bernini F, Paoletti R, Corsini A. Non-lipid-related effects of statins. *Ann Med*. 2000;32(3):164–176.
62. Sen-Banerjee S, Mir S, Lin Z, et al. Kruppel-like factor 2 as a novel mediator of statin effects in endothelial cells. *Circulation*. 2005;112(5):720–726.
63. van Thienen JV, Fledderus JO, Dekker RJ, et al. Shear stress sustains atheroprotective endothelial KLF2 expression more potently than statins through mRNA stabilization. *Cardiovasc Res*. 2006;72(2):231–240.
64. Rossi J, Rouleau L, Tardif JC, Leask RL. Effect of simvastatin on Kruppel-like factor2, endothelial nitric oxide synthase and thrombomodulin expression in endothelial cells under shear stress. *Life Sci*. 2010;87(3–4):92–99.
65. Tuomisto TT, Lumivuori H, Kansanen E, et al. Simvastatin has an anti-inflammatory effect on macrophages via upregulation of an atheroprotective transcription factor, Kruppel-like factor 2. *Cardiovasc Res*. 2008;78(1):175–184.
66. Bu DX, Tarrio M, Grabie N, et al. Statin-induced Kruppel-like factor 2 expression in human and mouse T cells reduces inflammatory and pathogenic responses. *J Clin Invest*. 2010;120(6):1961–1970.
67. Russo L, Gracia-Sancho J, Garcia-Caldero H, et al. Addition of simvastatin to cold storage solution prevents endothelial dysfunction in explanted rat livers. *Hepatology*. 2012;55(3):921–930.
68. Gracia-Sancho J, Garcia-Caldero H, Hide D, et al. Simvastatin maintains function and viability of steatotic rat livers procured for transplantation. *J Hepatol*. Jun 2013;58(6):1140–1146.
69. Yang L, Zhou X, Guo R, Shi Y, Liang X, Heng X. Role of Kruppel-like factor 2 and protease-activated receptor-1 in vulnerable plaques of ApoE(-/-) mice and intervention with statin. *Can J Cardiol*. 2013;29(8):997–1005.
70. Baur JA, Sinclair DA. Therapeutic potential of resveratrol: the in vivo evidence. *Nat Rev Drug Discov*. 2006;5(6):493–506.
71. Gracia-Sancho J, Villarreal Jr G, Zhang Y, Garcia-Cardena G. Activation of SIRT1 by resveratrol induces KLF2 expression conferring an endothelial vasoprotective phenotype. *Cardiovasc Res*. 2010;85(3):514–519.
72. Jain RK. Normalization of tumor vasculature: an emerging concept in antiangiogenic therapy. *Science*. 2005;307(5706):58–62.

73. Fukumura D, Duda DG, Munn LL, Jain RK. Tumor microvasculature and microenvironment: novel insights through intravital imaging in pre-clinical models. *Microcirculation.* 2010;17(3):206–225.
74. Hamzah J, Jugold M, Kiessling F, et al. Vascular normalization in Rgs5-deficient tumours promotes immune destruction. *Nature.* 2008;453(7193):410–414.
75. Hanahan D, Folkman J. Patterns and emerging mechanisms of the angiogenic switch during tumorigenesis. *Cell.* 1996;86(3):353–364.
76. Chakroborty D, Sarkar C, Yu H, et al. Dopamine stabilizes tumor blood vessels by up-regulating angiopoietin 1 expression in pericytes and Kruppel-like factor-2 expression in tumor endothelial cells. *Proc Natl Acad Sci USA.* 2011;108(51):20730–20735.
77. Luscher TF, Steffel J, Eberli FR, et al. Drug-eluting stent and coronary thrombosis: biological mechanisms and clinical implications. *Circulation.* 2007;115(8):1051–1058.
78. Muldowney 3rd JA, Stringham JR, Levy SE, et al. Antiproliferative agents alter vascular plasminogen activator inhibitor-1 expression: a potential prothrombotic mechanism of drug-eluting stents. *Arterioscler Thromb Vasc Biol.* 2007;27(2):400–406.
79. Jin HK, Ahn SH, Yoon JW, et al. Rapamycin down-regulates inducible nitric oxide synthase by inducing proteasomal degradation. *Biol Pharm Bull.* 2009;32(6):988–992.
80. Ma Q, Nie X, Yu M, et al. Rapamycin regulates the expression and activity of Kruppel-like transcription factor 2 in human umbilical vein endothelial cells. *PLoS One.* 2012;7(8):e43315.
81. Hay N, Sonenberg N. Upstream and downstream of mTOR. *Genes Dev.* 2004;18(16):1926–1945.
82. Nie XM, Su LX, Xu RX, Guo YL, Zhou YJ, Li JJ. Kruppel-like factor 2 might mediate the rapamycin-induced arterial thrombosis in vivo: implications for stent thrombosis in patients. *Chin Med J (Engl).* 2013;126(14):2636–2640.
83. SenBanerjee S, Lin Z, Atkins GB, et al. KLF2 is a novel transcriptional regulator of endothelial proinflammatory activation. *J Exp Med.* 2004;199(10):1305–1315.
84. Lin Z, Kumar A, SenBanerjee S, et al. Kruppel-like factor 2 (KLF2) regulates endothelial thrombotic function. *Circ Res.* 2005;96(5):e48–e57.
85. Li M, Wang X, Fu W, He S, Li D, Ke Q. CD4+CD25+Foxp3+ regulatory T cells protect endothelial function impaired by oxidized low density lipoprotein via the KLF-2 transcription factor. *Cell Physiol Biochem.* 2011;28(4):639–648.
86. Kumar A, Kumar S, Vikram A, et al. Histone and DNA methylation-mediated epigenetic downregulation of endothelial Kruppel-like factor 2 by low-density lipoprotein cholesterol. *Arterioscler Thromb Vasc Biol.* 2013;33(8):1936–1942.
87. Lee HY, Youn SW, Oh BH, Kim HS. Kruppel-like factor 2 suppression by high glucose as a possible mechanism of diabetic vasculopathy. *Korean Circ J.* 2012;42(4): 239–245.
88. Lee HY, Youn SW, Cho HJ, et al. FOXO1 impairs whereas statin protects endothelial function in diabetes through reciprocal regulation of Kruppel-like factor 2. *Cardiovasc Res.* 2013;97(1):143–152.
89. Migliaccio E, Giorgio M, Mele S, et al. The p66shc adaptor protein controls oxidative stress response and life span in mammals. *Nature.* 1999;402(6759):309–313.
90. Pacini S, Pellegrini M, Migliaccio E, et al. p66SHC promotes apoptosis and antagonizes mitogenic signaling in T cells. *Mol Cell Biol.* 2004;24(4):1747–1757.
91. Napoli C, Martin-Padura I, de Nigris F, et al. Deletion of the p66Shc longevity gene reduces systemic and tissue oxidative stress, vascular cell apoptosis, and early atherogenesis in mice fed a high-fat diet. *Proc Natl Acad Sci USA.* 2003;100(4):2112–2116.
92. De S, Razorenova O, McCabe NP, O'Toole T, Qin J, Byzova TV. VEGF-integrin interplay controls tumor growth and vascularization. *Proc Natl Acad Sci USA.* 2005;102(21):7589–7594.

93. Kumar A, Hoffman TA, Dericco J, Naqvi A, Jain MK, Irani K. Transcriptional repression of Kruppel like factor-2 by the adaptor protein p66shc. *FASEB J.* 2009;23(12):4344–4352.
94. Levine AJ. p53, the cellular gatekeeper for growth and division. *Cell.* 1997;88(3):323–331.
95. Kumar A, Kim CS, Hoffman TA, et al. p53 impairs endothelial function by transcriptionally repressing Kruppel-Like Factor 2. *Arterioscler Thromb Vasc Biol.* 2011;31(1):133–141.
96. Fledderus JO, Boon RA, Volger OL, et al. KLF2 primes the antioxidant transcription factor Nrf2 for activation in endothelial cells. *Arterioscler Thromb Vasc Biol.* 2008;28(7):1339–1346.
97. Atkins GB, Wang Y, Mahabeleshwar GH, et al. Hemizygous deficiency of Kruppel-like factor 2 augments experimental atherosclerosis. *Circ Res.* 2008;103(7):690–693.
98. Drexler H, Hornig B. Endothelial dysfunction in human disease. *J Mol Cell Cardiol.* 1999;31(1):51–60.
99. Malek AM, Greene AL, Izumo S. Regulation of endothelin 1 gene by fluid shear stress is transcriptionally mediated and independent of protein kinase C and cAMP. *Proc Natl Acad Sci USA.* 1993;90(13):5999–6003.
100. Chauhan SD, Nilsson H, Ahluwalia A, Hobbs AJ. Release of C-type natriuretic peptide accounts for the biological activity of endothelium-derived hyperpolarizing factor. *Proc Natl Acad Sci USA.* 2003;100(3):1426–1431.
101. Razani B, Engelman JA, Wang XB, et al. Caveolin-1 null mice are viable but show evidence of hyperproliferative and vascular abnormalities. *J Biol Chem.* 2001;276(41):38121–38138.
102. Kaminski A, Pohl CB, Sponholz C, et al. Up-regulation of endothelial nitric oxide synthase inhibits pulmonary leukocyte migration following lung ischemia-reperfusion in mice. *Am J Pathol.* 2004;164(6):2241–2249.
103. De Caterina R, Libby P, Peng HB, et al. Nitric oxide decreases cytokine-induced endothelial activation. Nitric oxide selectively reduces endothelial expression of adhesion molecules and proinflammatory cytokines. *J Clin Invest.* 1995;96(1):60–68.
104. Tziros C, Freedman JE. The many antithrombotic actions of nitric oxide. *Curr Drug Targets.* 2006;7(10):1243–1251.
105. Dekker RJ, Boon RA, Rondaij MG, et al. KLF2 provokes a gene expression pattern that establishes functional quiescent differentiation of the endothelium. *Blood.* 2006;107(11):4354–4363.
106. van Agtmaal EL, Bierings R, Dragt BS, et al. The shear stress-induced transcription factor KLF2 affects dynamics and angiopoietin-2 content of Weibel-Palade bodies. *PLoS One.* 2012;7(6):e38399.
107. Rand JH. The antiphospholipid syndrome. *Annu Rev Med.* 2003;54:409–424.
108. Simantov R, LaSala JM, Lo SK, et al. Activation of cultured vascular endothelial cells by antiphospholipid antibodies. *J Clin Invest.* 1995;96(5):2211–2219.
109. De Martin R, Hoeth M, Hofer-Warbinek R, Schmid JA. The transcription factor NF-kappa B and the regulation of vascular cell function. *Arterioscler Thromb Vasc Biol.* 2000;20(11):E83–E88.
110. Allen KL, Hamik A, Jain MK, McCrae KR. Endothelial cell activation by antiphospholipid antibodies is modulated by Kruppel-like transcription factors. *Blood.* 2011;117(23):6383–6391.
111. Lin Z, Hamik A, Jain R, Kumar A, Jain MK. Kruppel-like factor 2 inhibits protease activated receptor-1 expression and thrombin-mediated endothelial activation. *Arterioscler Thromb Vasc Biol.* 2006;26(5):1185–1189.

112. Helbing T, Rothweiler R, Ketterer E, et al. BMP activity controlled by BMPER regulates the proinflammatory phenotype of endothelium. *Blood.* 2011;118(18): 5040–5049.
113. Berk BC, Abe JI, Min W, Surapisitchat J, Yan C. Endothelial atheroprotective and anti-inflammatory mechanisms. *Ann N Y Acad Sci.* 2001;947:93–109, discussion 109–111.
114. Kracht M, Saklatvala J. Transcriptional and post-transcriptional control of gene expression in inflammation. *Cytokine.* 2002;20(3):91–106.
115. Fledderus JO, van Thienen JV, Boon RA, et al. Prolonged shear stress and KLF2 suppress constitutive proinflammatory transcription through inhibition of ATF2. *Blood.* 2007;109(10):4249–4257.
116. Boon RA, Leyen TA, Fontijn RD, et al. KLF2-induced actin shear fibers control both alignment to flow and JNK signaling in vascular endothelium. *Blood.* 2010;115(12):2533–2542.
117. Kaunas R, Usami S, Chien S. Regulation of stretch-induced JNK activation by stress fiber orientation. *Cell Signal.* 2006;18(11):1924–1931.
118. Shaik SS, Soltau TD, Chaturvedi G, et al. Low intensity shear stress increases endothelial ELR + CXC chemokine production via a focal adhesion kinase-p38{beta} MAPK-NF-{kappa} B pathway. *J Biol Chem.* 2009;284(9):5945–5955.
119. Boon RA, Fledderus JO, Volger OL, et al. KLF2 suppresses TGF-beta signaling in endothelium through induction of Smad7 and inhibition of AP-1. *Arterioscler Thromb Vasc Biol.* Mar 2007;27(3):532–539.
120. Egorova AD, Van der Heiden K, Van de Pas S, et al. Tgfbeta/Alk5 signaling is required for shear stress induced klf2 expression in embryonic endothelial cells. *Dev Dyn.* 2011;240(7):1670–1680.
121. Kinderlerer AR, Ali F, Johns M, et al. KLF2-dependent, shear stress-induced expression of CD59: a novel cytoprotective mechanism against complement-mediated injury in the vasculature. *J Biol Chem.* 2008;283(21):14636–14644.
122. Lin Z, Natesan V, Shi H, et al. Kruppel-like factor 2 regulates endothelial barrier function. *Arterioscler Thromb Vasc Biol.* 2010;30(10):1952–1959.
123. Shi H, Sheng B, Zhang F, et al. Kruppel-like factor 2 protects against ischemic stroke by regulating endothelial blood brain barrier function. *Am J Physiol Heart Circ Physiol.* 2013;304(6):H796–H805.
124. Wu C, Li F, Han G, Liu Z. Abeta(1-42) disrupts the expression and function of KLF2 in Alzheimer's disease mediated by p53. *Biochem Biophys Res Commun.* 2013;431(2):141–145.
125. Strazielle N, Ghersi-Egea JF, Ghiso J, et al. In vitro evidence that beta-amyloid peptide 1-40 diffuses across the blood-brain barrier and affects its permeability. *J Neuropathol Exp Neurol.* 2000;59(1):29–38.
126. McGuinness B, Passmore P. Can statins prevent or help treat Alzheimer's disease? *J Alzheimers Dis.* 2010;20(3):925–933.
127. Kim DW, Langille BL, Wong MK, Gotlieb AI. Patterns of endothelial microfilament distribution in the rabbit aorta in situ. *Circ Res.* 1989;64(1):21–31.
128. Davies PF, Barbee KA, Volin MV, et al. Spatial relationships in early signaling events of flow-mediated endothelial mechanotransduction. *Annu Rev Physiol.* 1997;59:527–549.
129. Wong CW, Christen T, Roth I, et al. Connexin37 protects against atherosclerosis by regulating monocyte adhesion. *Nat Med.* 2006;12(8):950–954.
130. Kwak BR, Mulhaupt F, Veillard N, Gros DB, Mach F. Altered pattern of vascular connexin expression in atherosclerotic plaques. *Arterioscler Thromb Vasc Biol.* 2002;22(2):225–230.

131. Saez JC, Berthoud VM, Branes MC, Martinez AD, Beyer EC. Plasma membrane channels formed by connexins: their regulation and functions. *Physiol Rev.* 2003;83(4): 1359–1400.
132. Pfenniger A, Wong C, Sutter E, et al. Shear stress modulates the expression of the atheroprotective protein Cx37 in endothelial cells. *J Mol Cell Cardiol.* 2012;53(2):299–309.
133. Arai M, Ohashi T, Tsukahara T, et al. Human T-cell leukemia virus type 1 Tax protein induces the expression of lymphocyte chemoattractant SDF-1/PBSF. *Virology.* 1998;241(2):298–303.
134. Ryter SW, Alam J, Choi AM. Heme oxygenase-1/carbon monoxide: from basic science to therapeutic applications. *Physiol Rev.* 2006;86(2):583–650.
135. Brunt KR, Fenrich KK, Kiani G, et al. Protection of human vascular smooth muscle cells from H2O2-induced apoptosis through functional codependence between HO-1 and AKT. *Arterioscler Thromb Vasc Biol.* 2006;26(9):2027–2034.
136. Otterbein LE, Zuckerbraun BS, Haga M, et al. Carbon monoxide suppresses arteriosclerotic lesions associated with chronic graft rejection and with balloon injury. *Nat Med.* 2003;9(2):183–190.
137. Brouard S, Otterbein LE, Anrather J, et al. Carbon monoxide generated by heme oxygenase 1 suppresses endothelial cell apoptosis. *J Exp Med.* 2000;192(7):1015–1026.
138. Ali F, Hamdulay SS, Kinderlerer AR, et al. Statin-mediated cytoprotection of human vascular endothelial cells: a role for Kruppel-like factor 2-dependent induction of heme oxygenase-1. *J Thromb Haemost.* 2007;5(12):2537–2546.
139. Hergenreider E, Heydt S, Treguer K, et al. Atheroprotective communication between endothelial cells and smooth muscle cells through miRNAs. *Nat Cell Biol.* 2012;14(3):249–256.
140. Kuo CT, Veselits ML, Barton KP, Lu MM, Clendenin C, Leiden JM. The LKLF transcription factor is required for normal tunica media formation and blood vessel stabilization during murine embryogenesis. *Genes Dev.* 1997;11(22):2996–3006.
141. Wu J, Bohanan CS, Neumann JC, Lingrel JB. KLF2 transcription factor modulates blood vessel maturation through smooth muscle cell migration. *J Biol Chem.* 2008;283(7):3942–3950.
142. Perry HM, Bender TP, McNamara CA. B cell subsets in atherosclerosis. *Front Immunol.* 2012;3:373.
143. Mallat Z, Taleb S, Ait-Oufella H, Tedgui A. The role of adaptive T cell immunity in atherosclerosis. *J Lipid Res.* 2009;50(suppl):S364–S369.
144. Buckley AF, Kuo CT, Leiden JM. Transcription factor LKLF is sufficient to program T cell quiescence via a c-Myc-dependent pathway. *Nat Immunol.* 2001;2(8):698–704.
145. Kyaw T, Winship A, Tay C, et al. Cytotoxic and proinflammatory CD8+ T lymphocytes promote development of vulnerable atherosclerotic plaques in apoE-deficient mice. *Circulation.* 2013;127(9):1028–1039.
146. Carlson CM, Endrizzi BT, Wu J, et al. Kruppel-like factor 2 regulates thymocyte and T-cell migration. *Nature.* 2006;442(7100):299–302.
147. Ait-Oufella H, Salomon BL, Potteaux S, et al. Natural regulatory T cells control the development of atherosclerosis in mice. *Nat Med.* 2006;12(2):178–180.
148. Mor A, Planer D, Luboshits G, et al. Role of naturally occurring CD4+ CD25+ regulatory T cells in experimental atherosclerosis. *Arterioscler Thromb Vasc Biol.* 2007;27(4):893–900.
149. Kyaw T, Tay C, Krishnamurthi S, et al. B1a B lymphocytes are atheroprotective by secreting natural IgM that increases IgM deposits and reduces necrotic cores in atherosclerotic lesions. *Circ Res.* 2011;109(8):830–840.
150. Libby P. Inflammation in atherosclerosis. *Nature.* 2002;420(6917):868–874.

151. Lingrel JB, Pilcher-Roberts R, Basford JE, et al. Myeloid-specific Kruppel-like factor 2 inactivation increases macrophage and neutrophil adhesion and promotes atherosclerosis. *Circ Res*. 2012;110(10):1294–1302.
152. Mantovani A, Garlanda C, Locati M. Macrophage diversity and polarization in atherosclerosis: a question of balance. *Arterioscler Thromb Vasc Biol*. 2009;29(10): 1419–1423.
153. Goerdt S, Orfanos CE. Other functions, other genes: alternative activation of antigen-presenting cells. *Immunity*. 1999;10(2):137–142.
154. Peiser L, Gordon S. The function of scavenger receptors expressed by macrophages and their role in the regulation of inflammation. *Microbes Infect*. 2001;3(2):149–159.
155. van Tits LJ, Stienstra R, van Lent PL, Netea MG, Joosten LA, Stalenhoef AF. Oxidized LDL enhances pro-inflammatory responses of alternatively activated M2 macrophages: a crucial role for Kruppel-like factor 2. *Atherosclerosis*. 2011;214(2):345–349.
156. Wani MA, Means Jr RT, Lingrel JB. Loss of LKLF function results in embryonic lethality in mice. *Transgenic Res*. 1998;7(4):229–238.
157. Leung DW, Cachianes G, Kuang WJ, Goeddel DV, Ferrara N. Vascular endothelial growth factor is a secreted angiogenic mitogen. *Science*. 1989;246(4935):1306–1309.
158. Senger DR, Galli SJ, Dvorak AM, Perruzzi CA, Harvey VS, Dvorak HF. Tumor cells secrete a vascular permeability factor that promotes accumulation of ascites fluid. *Science*. 1983;219(4587):983–985.
159. Ferrara N, Gerber HP, LeCouter J. The biology of VEGF and its receptors. *Nat Med*. 2003;9(6):669–676.
160. Maharaj AS, D'Amore PA. Roles for VEGF in the adult. *Microvasc Res*. 2007;74(2–3):100–113.
161. Kim I, Moon SO, Kim SH, Kim HJ, Koh YS, Koh GY. Vascular endothelial growth factor expression of intercellular adhesion molecule 1 (ICAM-1), vascular cell adhesion molecule 1 (VCAM-1), and E-selectin through nuclear factor-kappa B activation in endothelial cells. *J Biol Chem*. 2001;276(10):7614–7620.
162. Waltenberger J, Claesson-Welsh L, Siegbahn A, Shibuya M, Heldin CH. Different signal transduction properties of KDR and Flt1, two receptors for vascular endothelial growth factor. *J Biol Chem*. 1994;269(43):26988–26995.
163. Nicoli S, Standley C, Walker P, Hurlstone A, Fogarty KE, Lawson ND. MicroRNA-mediated integration of haemodynamics and Vegf signalling during angiogenesis. *Nature*. 2010;464(7292):1196–1200.
164. Meadows SM, Salanga MC, Krieg PA. Kruppel-like factor 2 cooperates with the ETS family protein ERG to activate Flk1 expression during vascular development. *Development*. 2009;136(7):1115–1125.
165. dela Paz NG, Walshe TE, Leach LL, Saint-Geniez M, D'Amore PA. Role of shear-stress-induced VEGF expression in endothelial cell survival. *J Cell Sci*. 2012;125(Pt 4):831–843.
166. Bhattacharya R, Senbanerjee S, Lin Z, et al. Inhibition of vascular permeability factor/vascular endothelial growth factor-mediated angiogenesis by the Kruppel-like factor KLF2. *J Biol Chem*. 2005;280(32):28848–28851.
167. Kawanami D, Mahabeleshwar GH, Lin Z, et al. Kruppel-like factor 2 inhibits hypoxia-inducible factor 1alpha expression and function in the endothelium. *J Biol Chem*. 2009;284(31):20522–20530.
168. Wong AL, Haroon ZA, Werner S, Dewhirst MW, Greenberg CS, Peters KG. Tie2 expression and phosphorylation in angiogenic and quiescent adult tissues. *Circ Res*. 1997;81(4):567–574.
169. Peters KG, Kontos CD, Lin PC, et al. Functional significance of Tie2 signaling in the adult vasculature. *Recent Prog Horm Res*. 2004;59:51–71.

170. Fukuhara S, Sako K, Minami T, et al. Differential function of Tie2 at cell-cell contacts and cell-substratum contacts regulated by angiopoietin-1. *Nat Cell Biol.* 2008;10(5):513–526.
171. Saharinen P, Eklund L, Miettinen J, et al. Angiopoietins assemble distinct Tie2 signalling complexes in endothelial cell-cell and cell-matrix contacts. *Nat Cell Biol.* 2008;10(5):527–537.
172. Sako K, Fukuhara S, Minami T, et al. Angiopoietin-1 induces Kruppel-like factor 2 expression through a phosphoinositide 3-kinase/AKT-dependent activation of myocyte enhancer factor 2. *J Biol Chem.* 2009;284(9):5592–5601.
173. Gu H, Neel BG. The "Gab" in signal transduction. *Trends Cell Biol.* 2003;13(3):122–130.
174. Shioyama W, Nakaoka Y, Higuchi K, et al. Docking protein Gab1 is an essential component of postnatal angiogenesis after ischemia via HGF/c-met signaling. *Circ Res.* 2011;108(6):664–675.
175. Higuchi K, Nakaoka Y, Shioyama W, et al. Endothelial Gab1 deletion accelerates angiotensin II-dependent vascular inflammation and atherosclerosis in apolipoprotein E knockout mice. *Circ J.* 2012;76(8):2031–2040.
176. Asahara T, Murohara T, Sullivan A, et al. Isolation of putative progenitor endothelial cells for angiogenesis. *Science.* 1997;275(5302):964–967.
177. Vasa M, Fichtlscherer S, Aicher A, et al. Number and migratory activity of circulating endothelial progenitor cells inversely correlate with risk factors for coronary artery disease. *Circ Res.* 2001;89(1):E1–E7.
178. Boon RA, Urbich C, Fischer A, et al. Kruppel-like factor 2 improves neovascularization capacity of aged proangiogenic cells. *Eur Heart J.* 2011;32(3):371–377.
179. Egorova AD, DeRuiter MC, de Boer HC, et al. Endothelial colony-forming cells show a mature transcriptional response to shear stress. *In Vitro Cell Dev Biol Anim.* 2012;48(1):21–29.
180. Armstrong EJ, Bischoff J. Heart valve development: endothelial cell signaling and differentiation. *Circ Res.* 2004;95(5):459–470.
181. Vermot J, Forouhar AS, Liebling M, et al. Reversing blood flows act through klf2a to ensure normal valvulogenesis in the developing heart. *PLoS Biol.* 2009;7(11):e1000246.
182. Just S, Berger IM, Meder B, et al. Protein kinase D2 controls cardiac valve formation in zebrafish by regulating histone deacetylase 5 activity. *Circulation.* 2011;124(3):324–334.
183. Chiplunkar AR, Lung TK, Alhashem Y, et al. Kruppel-like factor 2 is required for normal mouse cardiac development. *PLoS One.* 2013;8(2):e54891.

# INDEX

Note: Page numbers followed by "*f*" indicate figures and "*t*" indicate tables.

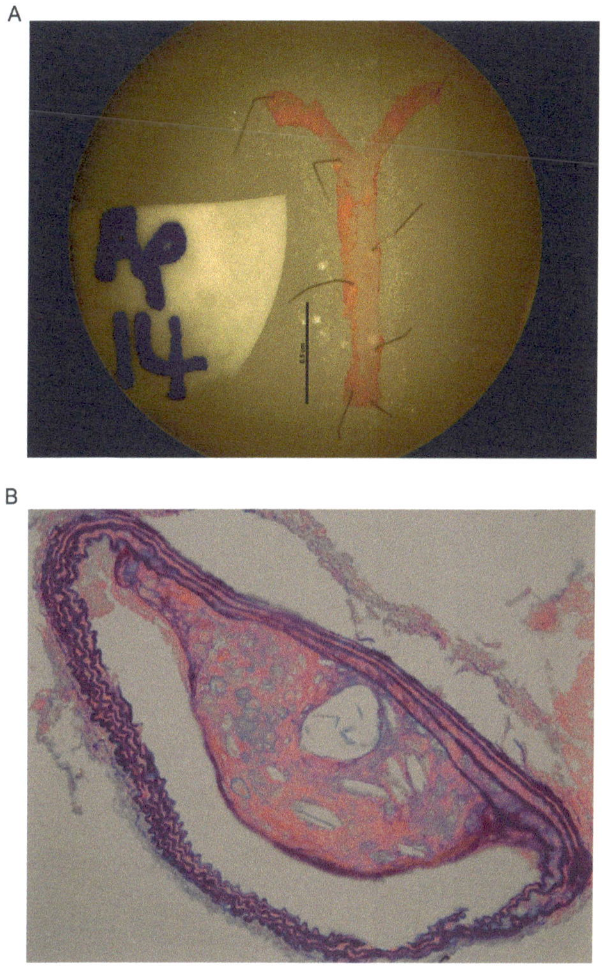


SHEILA E. FRANCIS, FIGURE 1.1

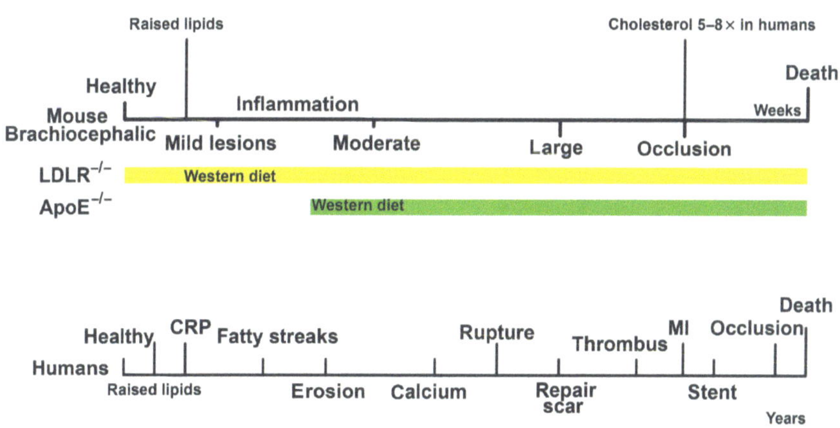


SHEILA E. FRANCIS, FIGURE 1.2

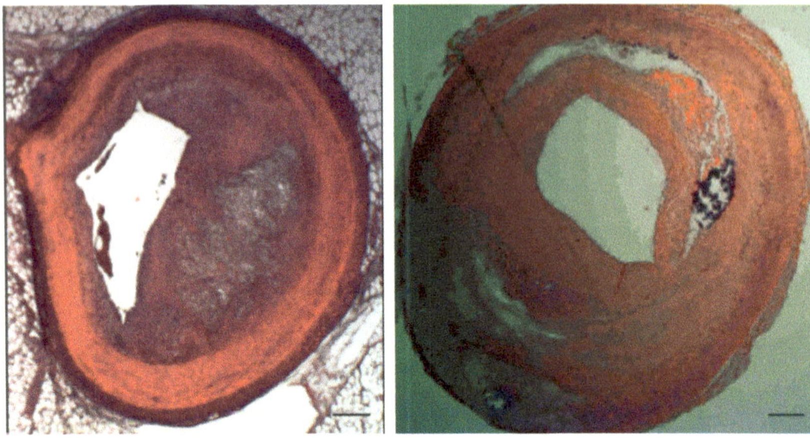

Sheila E. Francis, Figure 1.3

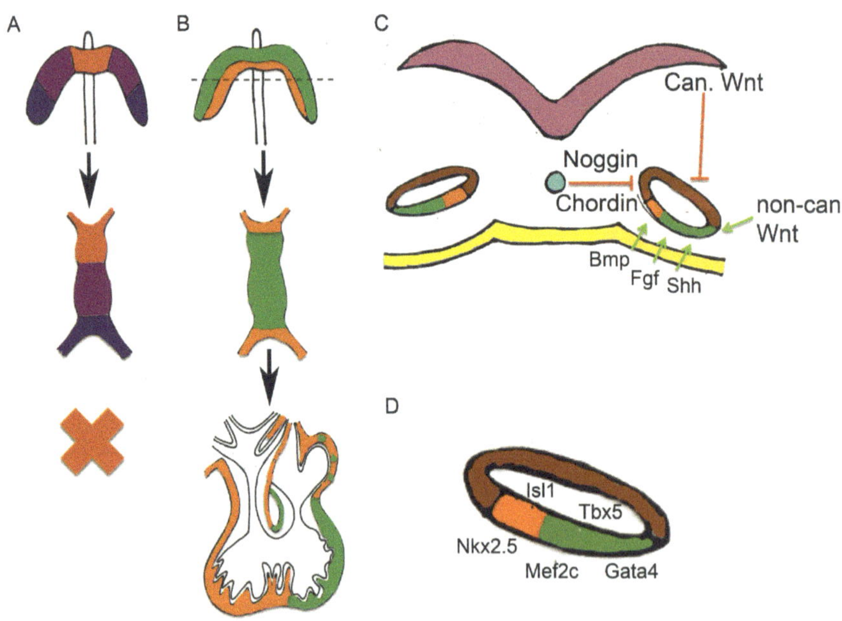


Bill Chaudhry *et al.*, Figure 2.1

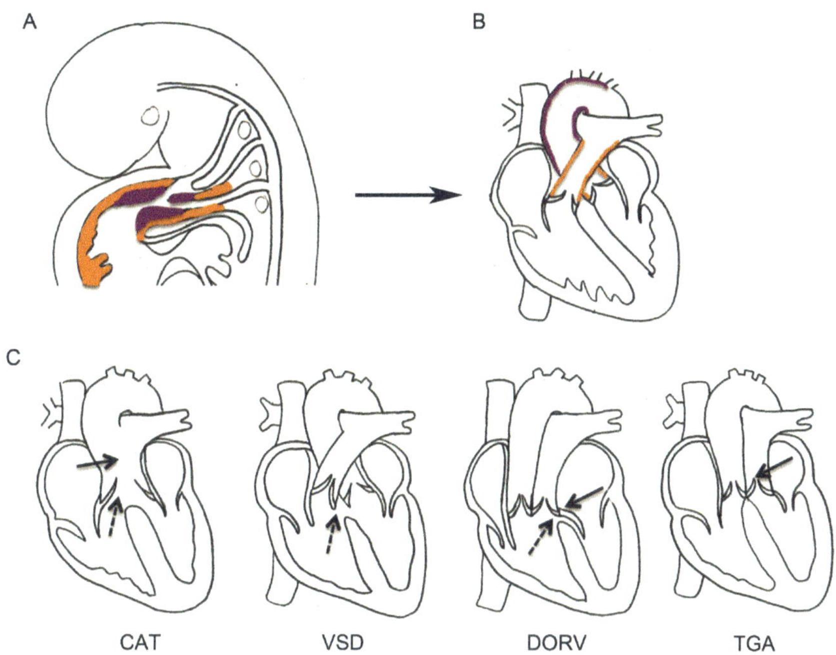


Bill Chaudhry *et al.*, Figure 2.2

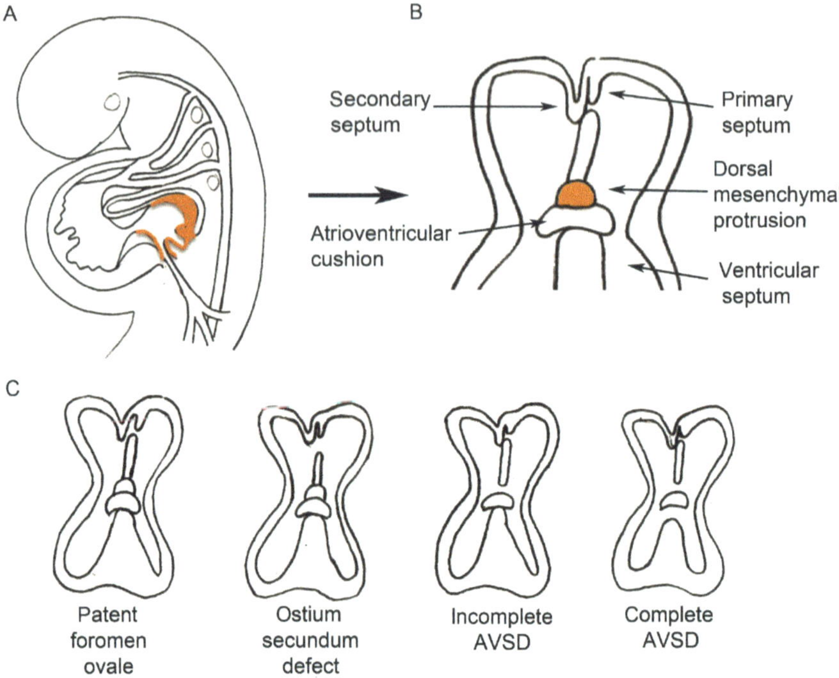


Bill Chaudhry *et al.*, Figure 2.3

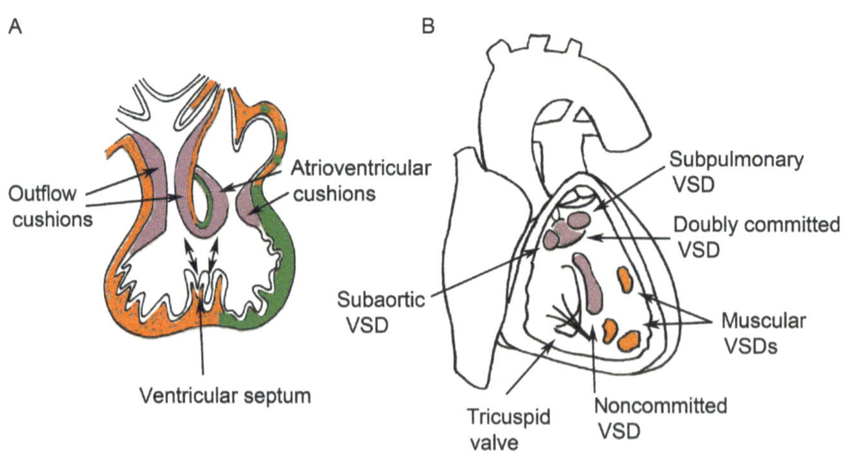


Bill Chaudhry *et al.*, Figure 2.4

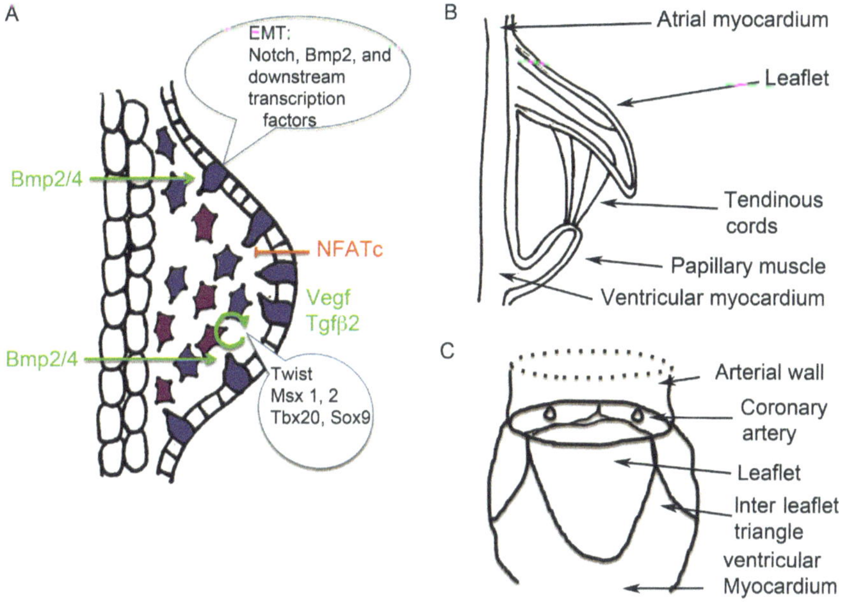


Bill Chaudhry *et al.*, Figure 2.5

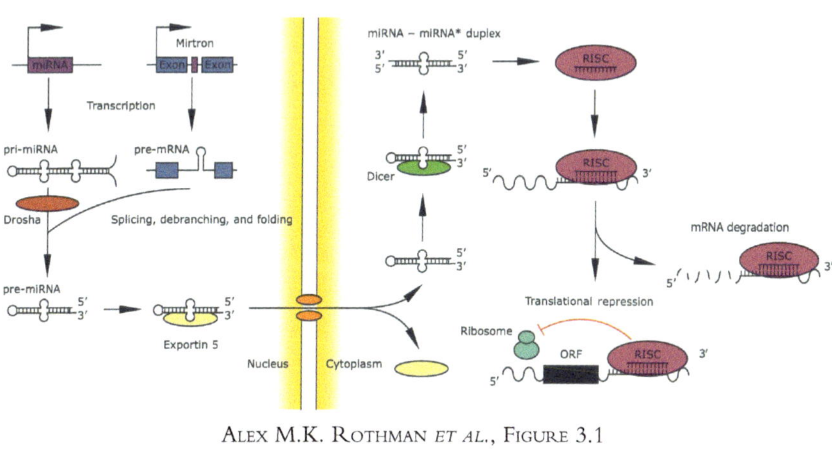


Alex M.K. Rothman *et al.*, Figure 3.1

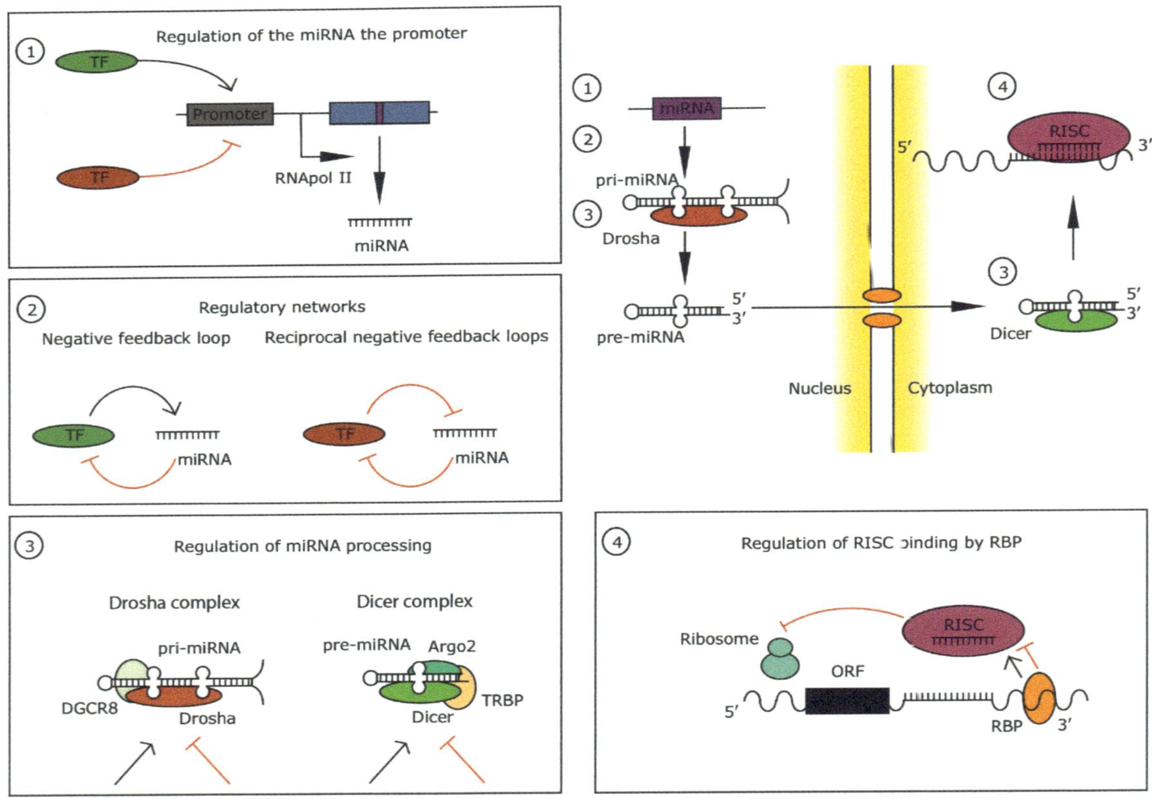


ALEX M.K. ROTHMAN *ET AL.*, FIGURE 3.2

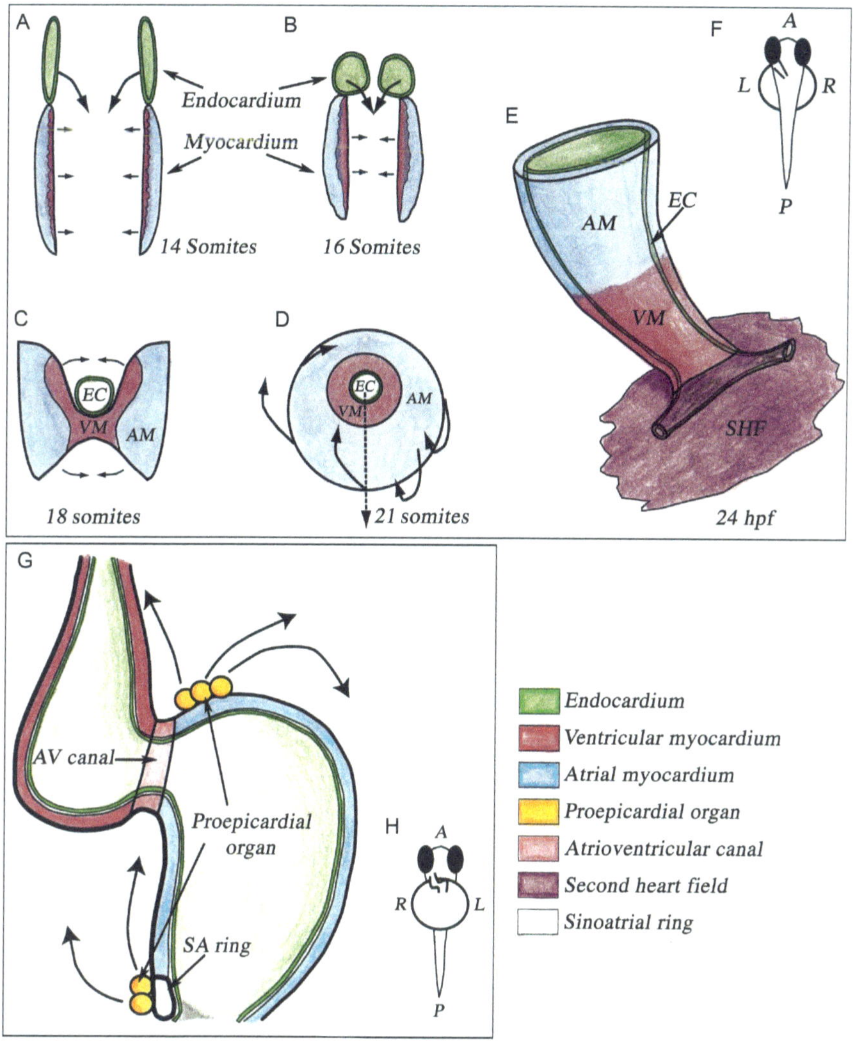


Robert N. Wilkinson *et al.*, Figure 4.1

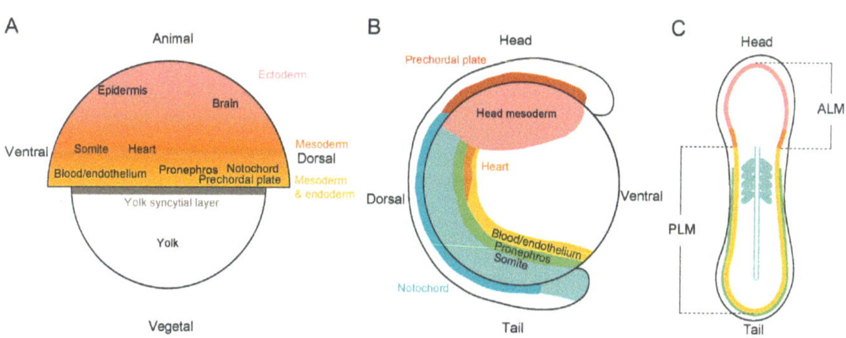


Robert N. Wilkinson and Fredericus J.M. van Eeden, Figure 5.1

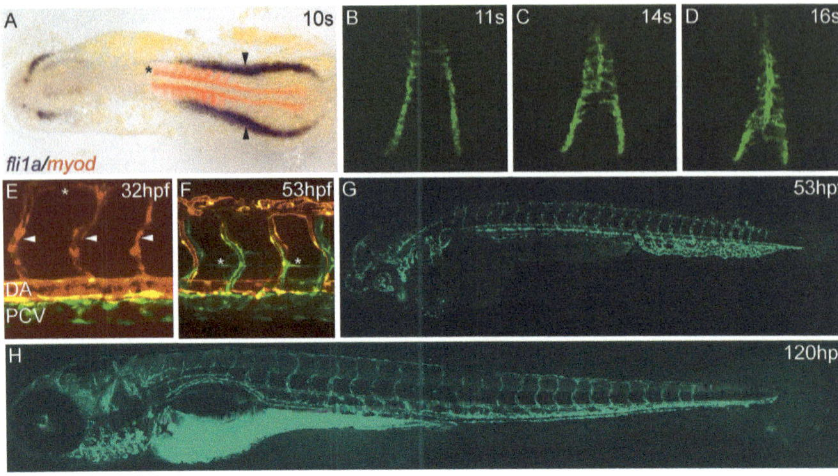


ROBERT N. WILKINSON AND FREDERICUS J.M. VAN EEDEN, FIGURE 5.2

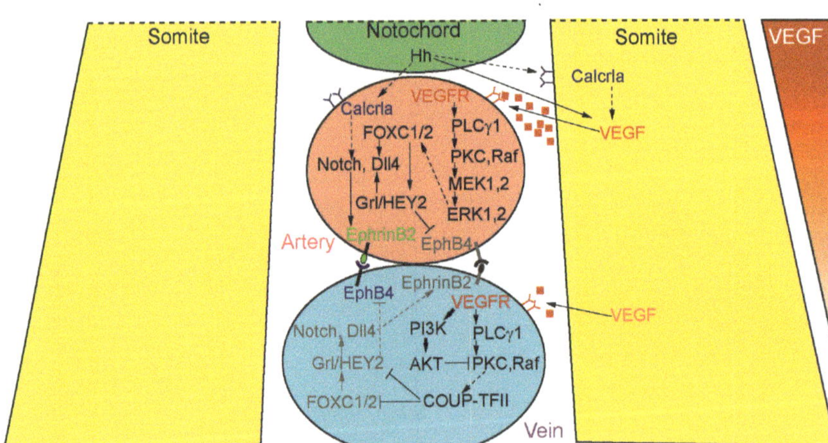


ROBERT N. WILKINSON AND FREDERICUS J.M. VAN EEDEN, FIGURE 5.3

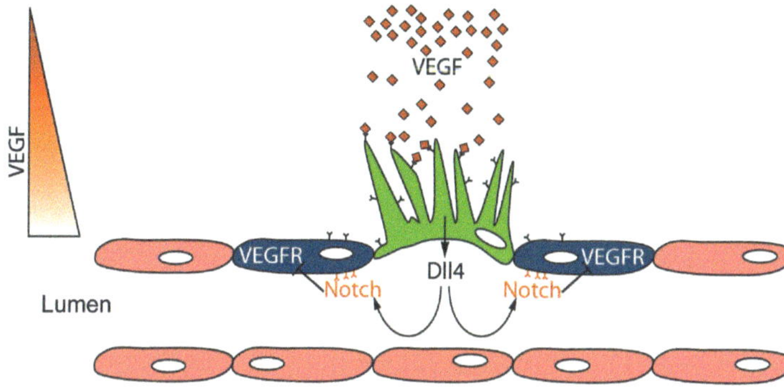


Robert N. Wilkinson and Fredericus J.M. van Eeden, Figure 5.4

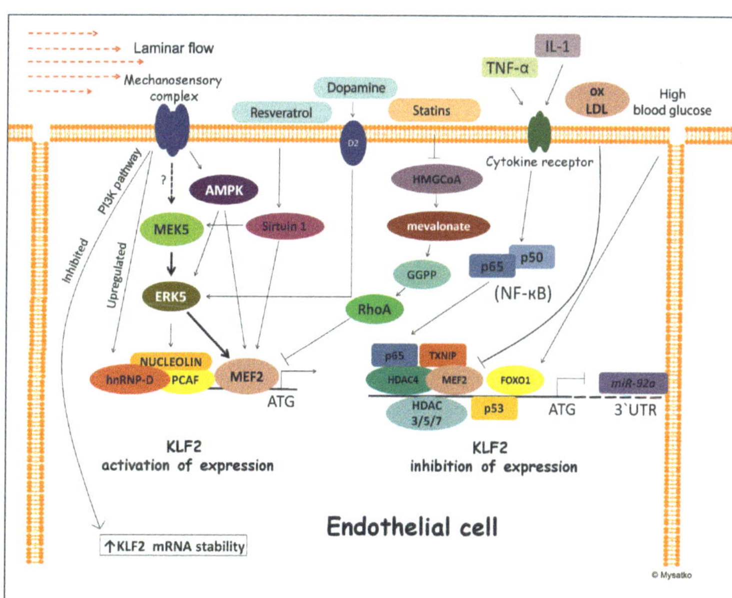


Peter Novodvorsky and Timothy J.A. Chico, Figure 7.1

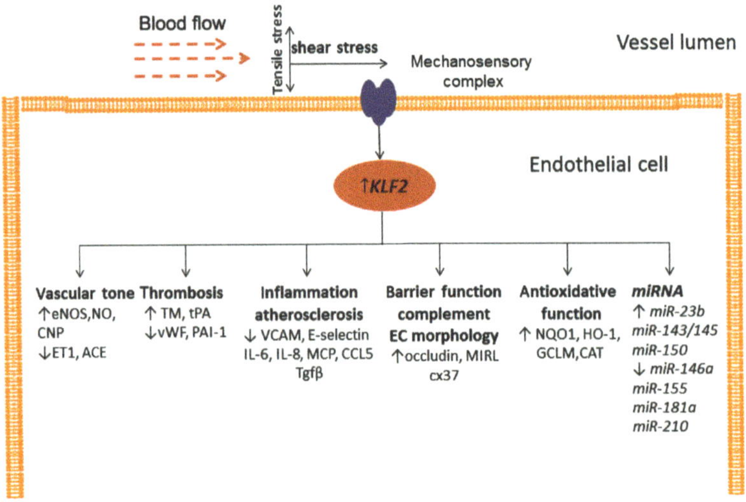


Peter Novodvorsky and Timothy J.A. Chico, Figure 7.2

FSC
www.fsc.org
MIX
Papier | Fördert gute Waldnutzung
FSC® C083411

Kolibri 360 GmbH
Ferdinand-Jühlke-Straße 7
99095 Erfurt, Deutschland
produktsicherheit@kolibri360.de